T0189336

Studies in Computational Intelligence

Volume 844

The series "Studies in Computational Intelligence" (SCI) publishes new developments and advances in the various areas of computational intelligence—quickly and with a high quality. The intent is to cover the theory, applications, and design methods of computational intelligence, as embedded in the fields of engineering, computer science, physics and life sciences, as well as the methodologies behind them. The series contains monographs, lecture notes and edited volumes in computational intelligence spanning the areas of neural networks, connectionist systems, genetic algorithms, evolutionary computation, artificial intelligence, cellular automata, self-organizing systems, soft computing, fuzzy systems, and hybrid intelligent systems. Of particular value to both the contributors and the readership are the short publication timeframe and the world-wide distribution, which enable both wide and rapid dissemination of research output.

The books of this series are submitted to indexing to Web of Science, EI-Compendex, DBLP, SCOPUS, Google Scholar and Springerlink.

More information about this series at http://www.springer.com/series/7092

Roger Lee
Editor

Big Data, Cloud Computing, and Data Science Engineering

Editor
Roger Lee
Software Engineering and Information
Technology Institute
Central Michigan University
Mount Pleasant, MI, USA

ISSN 1860-949X ISSN 1860-9503 (electronic)
Studies in Computational Intelligence
ISBN 978-3-030-24407-1 ISBN 978-3-030-24405-7 (eBook)
https://doi.org/10.1007/978-3-030-24405-7

This Springer imprint is published by the registered company Springer Nature Switzerland AG
The registered company address is: Gewerbestrasse 11, 6330 Cham, Switzerland

Foreword

The purpose of the 4th IEEE/ACIS International Conference on Big Data, Cloud Computing, Data Science and Engineering (BCD) held on May 29–31, 2019 in Honolulu, Hawaii was for researchers, scientists, engineers, industry practitioners, and students to discuss, encourage and exchange new ideas, research results, and experiences on all aspects of Applied Computers and Information Technology, and to discuss the practical challenges encountered along the way and the solutions adopted to solve them. The conference organizers have selected the best 13 papers from those papers accepted for presentation at the conference in order to publish them in this volume. The papers were chosen based on review scores submitted by members of the program committee and underwent further rigorous rounds of review.

In chapter "Robust Optimization Model for Designing Emerging Cloud-Fog Networks", Masayuki Tsujino proposes a robust design model for economically constructing IoT infrastructures. They experimentally evaluated the effectiveness of the proposed model and the possibility of applying the method to this model to practical scaled networks.

In chapter "Multi-task Deep Reinforcement Learning with Evolutionary Algorithm and Policy Gradients Method in 3D Control Tasks", Shota Imai, Yuichi Sei, Yasuyuki Tahara, Ryohei Orihara, and Akihiko Ohsuga propose a pretraining method to train a model that can work well on variety of target tasks and solve the problems with deep reinforcement learning with an evolutionary algorithm and policy gradients method. In this method, agents explore multiple environments with a diverse set of neural networks to train a general model with evolutionary algorithm and policy gradients method.

In chapter "Learning Neural Circuit by AC Operation and Frequency Signal Output", Masashi Kawaguchi, Naohiro Ishii, and Masayoshi Umeno used analog electronic circuits using alternating current to realize the neural network learning model. These circuits are composed of a rectifier circuit, voltage-frequency converter, amplifier, subtract circuit, additional circuit, and inverter. They suggest the realization of the deep learning model regarding the proposed analog hardware neural circuit.

In chapter "IoTDoc: A Docker-Container Based Architecture of IoT-Enabled Cloud System", Shahid Noor, Bridget Koehler, Abby Steenson, Jesus Caballero, David Ellenberger, and Lucas Heilman introduce IoTDoc, an architecture of mobile cloud composed of lightweight containers running on distributed IoT devices. To explore the benefits of running containers on low-cost IoT-based cloud system, they use Docker to create and orchestrate containers and run on a cloud formed by cluster of IoT devices. Their experimental result shows that IoTDoc is a viable option for cloud computing and is a more affordable, cost-effective alternative to large platform cloud computing services.

In chapter "A Survival Analysis-Based Prioritization of Code Checker Warning: A Case Study Using PMD", Hirohisa Aman, Sousuke Amasaki, Tomoyuki Yokogawa, and Minoru Kawahara propose an application of the survival analysis method to prioritize code checker warnings. The proposed method estimates a warning's lifetime with using the real trend of warnings through code changes; the brevity of warning means its importance because severe warnings are related to problematic parts which programmers would fix sooner.

In chapter "Elevator Monitoring System to Guide User's Behavior by Visualizing the State of Crowdedness", Haruhisa Hasegawa and Shiori Aida propose that even old equipment can be made efficient using IoT. They propose an IoT system that improves the fairness and efficiency by visualizing the crowdedness of an elevator, which has only one cage. When a certain floor gets crowded, unfairness arises in the users on the other floors as they are not able to take the elevator. Their proposed system improves the fairness and efficiency by guiding the user's behavior.

In chapter "Choice Behavior Analysis of Internet Access Services Using Supervised Learning Models", Ken Nishimatsu, Akiya Inoue, Miiru Saito, and Motoi Iwashita conduct a study to try to understand the Internet-access service choice behavior considering the current market in Japan. They propose supervised learning models to create differential descriptions of these user segments from the viewpoints of decision-making factors. The characteristics of these user segments are shown by using the estimated models.

In chapter "Norm-referenced Criteria for Strength of the Upper Limbs for the Korean High School Baseball Players Using Computer Assisted Isokinetic Equipment", Su-Hyun Kim and Jin-Wook Lee conducted a study to set the norm-referenced criteria for isokinetic muscular strength of the upper limbs (elbow and shoulder joint) for the Korean 83 high school baseball players. The provided criteria of peak torque and peak torque per body weight, set through the computer isokinetic equipment, are very useful information for high school baseball player, baseball coach, athletic trainer, and sports injury rehabilitation specialists in injury recovery and return to rehabilitation, to utilize as an objective clinical assessment data.

In chapter "A Feature Point Extraction and Comparison Method Through Representative Frame Extraction and Distortion Correction for 360° Realistic Contents", Byeongchan Park, Youngmo Kim, Seok-Yoon Kim propose a feature point extraction and similarity comparison method for 360° realistic images by extracting representative frames and correcting distortions. The proposed method is shown, through the experiments, to be superior in speed for the image comparison than other methods, and it is also advantageous when the data to be stored in the server increase in the future.

In chapter "Dimension Reduction by Word Clustering with Semantic Distance", Toshinori Deguchi and Naohiro Ishii propose a method of clustering words using the semantic distances of words, the dimension of document vectors is reduced to the number of word clusters. Word distance is able to be calculated by using WordNet. This method is free from the amount of words and documents. For especially small documents, they use word's definition in a dictionary and calculate the similarities between documents.

In chapter "Word-Emotion Lexicon for Myanmar Language", Thiri Marlar Swe and Phyu Hninn Myint describe the creation of Myanmar word-emotion lexicon, M-Lexicon, which contains six basic emotions: happiness, sadness, fear, anger, surprise, and disgust. Matrices, Term-Frequency Inversed Document Frequency (TF-IDF), and unity-based normalization are used in lexicon creation. Experiment shows that the M-Lexicon creation contains over 70% of correctly associated with six basic emotions.

In chapter "Release from the Curse of High Dimensional Data Analysis", Shuichi Shinmura proposes a solution to the curse of high dimensional data analysis. In this research, they introduce the reason why no researchers could succeed in the cancer gene diagnosis by microarrays from 1970.

In chapter "Evaluation of Inertial Sensor Configurations for Wearable Gait Analysis", Hongyu Zhao, Zhelong Wang, Sen Qiu, Jie Li, Fengshan Gao, and Jianjun Wang address the problem of detecting gait events based on inertial sensors and body sensor networks (BSNs). Experimental results show that angular rate holds the most reliable information for gait recognition during forward walking on level ground.

It is our sincere hope that this volume provides stimulation and inspiration, and that it will be used as a foundation for works to come.

May 2019

Atsushi Shimoda
Chiba Institute of Technology
Narashino, Japan

Prajak Chertchom
Thai-Nichi Institute of Technology
Bangkok, Thailand

BCD 2019 Program Co-chairs

Contents

Contributors

Shiori Aida Department of Mathematical and Physical Sciences, Japan Women's University, Tokyo, Japan

Hirohisa Aman Center for Information Technology, Ehime University, Matsuyama, Ehime, Japan

Sousuke Amasaki Faculty of Computer Science and Systems Engineering, Okayama Prefectural University, Soja, Okayama, Japan

Jesus Caballero St. Olaf College, Northfield, USA

Toshinori Deguchi National Institute of Technology, Gifu College, Gifu, Japan

David Ellenberger St. Olaf College, Northfield, USA

Fengshan Gao Department of Physical Education, Dalian University of Technology, Dalian, China

Haruhisa Hasegawa Department of Mathematical and Physical Sciences, Japan Women's University, Tokyo, Japan

Lucas Heilman St. Olaf College, Northfield, USA

Shota Imai The University of Electro-Communications, Tokyo, Japan

Akiya Inoue Chiba Institute of Technology, Narashino, Japan

Naohiro Ishii Department of Information Science, Aichi Institute of Technology, Toyota, Japan

Motoi Iwashita Chiba Institute of Technology, Narashino, Japan

Masashi Kawaguchi Department of Electrical & Electronic Engineering, Suzuka National College of Technology, Suzuka Mie, Japan

Minoru Kawahara Center for Information Technology, Ehime University, Matsuyama, Ehime, Japan

Seok-Yoon Kim Department of Computer Science and Engineering, Soongsil University, Seoul, Republic of Korea

Su-Hyun Kim Department of Sports Medicine, Affiliation Sunsoochon Hospital, Songpa-gu, Seoul, Republic of Korea

Youngmo Kim Department of Computer Science and Engineering, Soongsil University, Seoul, Republic of Korea

Bridget Koehler St. Olaf College, Northfield, USA

Jin-Wook Lee Department of Exercise Prescription and Rehabilitation, Dankook University, Cheonan-si, Chungcheongnam-do, Republic of Korea

Jie Li School of Control Science and Engineering, Dalian University of Technology, Dalian, China

Phyu Hninn Myint University of Computer Studies, Mandalay, Myanmar

Ken Nishimatsu NTT Network Technology Laboratories, NTT Corporation, Musashino, Japan

Shahid Noor Northern Kentucky University, Highland Heights, USA

Akihiko Ohsuga The University of Electro-Communications, Tokyo, Japan

Ryohei Orihara The University of Electro-Communications, Tokyo, Japan

Byeongchan Park Department of Computer Science and Engineering, Soongsil University, Seoul, Republic of Korea

Sen Qiu School of Control Science and Engineering, Dalian University of Technology, Dalian, China

Miiru Saito Chiba Institute of Technology, Narashino, Japan

Yuichi Sei The University of Electro-Communications, Tokyo, Japan

Shuichi Shinmura Emeritus Seikei University, Chiba, Japan

Abby Steenson St. Olaf College, Northfield, USA

Thiri Marlar Swe University of Computer Studies, Mandalay, Myanmar

Yasuyuki Tahara The University of Electro-Communications, Tokyo, Japan

Masayuki Tsujino NTT Network Technology Laboratories, Musashino-Shi, Tokyo, Japan

Masayoshi Umeno Department of Electronic Engineering, Chubu University, Kasugai, Aichi, Japan

Jianjun Wang Beijing Institute of Spacecraft System Engineering, Beijing, China

Zhelong Wang School of Control Science and Engineering, Dalian University of Technology, Dalian, China

Tomoyuki Yokogawa Faculty of Computer Science and Systems Engineering, Okayama Prefectural University, Soja, Okayama, Japan

Hongyu Zhao School of Control Science and Engineering, Dalian University of Technology, Dalian, China

Robust Optimization Model for Designing Emerging Cloud-Fog Networks

Masayuki Tsujino

Abstract I focus on designing the placement and capacity for Internet of Things (IoT) infrastructures consisting of three layers; cloud, fog, and communication. It is extremely difficult to predict the future demand of innovative IoT services; thus, I propose a robust design model for economically constructing IoT infrastructures under uncertain demands, which is formulated as a robust optimization problem. I also present a method of solving this problem, which is practically difficult to solve. I experimentally evaluated the effectiveness of the proposed model and the possibility of applying the method to this model to practical scaled networks.

1 Introduction

Internet of Things (IoT) has attracted a great deal of attention in various industrial fields as an enabler for Internet evolution [10, 23]. IoT devices enable us to improve advanced services in conjunction with remote computing, which is usually provided in clouds, and are expected to be applied to delay-sensitive services, such as automobile control, industrial machine control, and telemedicine, under the coming 5G high speed wireless access environment. Therefore, it is necessary to reduce communication delay with a cloud computer located in a data center in a backbone network. Fog-computing technologies, which have gained attention in recent years, satisfy this requirement by installing a computing function for processing workloads near user devices [7, 18, 19].

Major telecom carriers, such as Tier-1 ISPs, own data centers where cloud computers are located, remote sites where fog computers can be located, as well as communication infrastructures. By deploying fog computers to these remote sites, they will have the opportunity to use network function virtualization/software defined network (NFV/SDN) technologies to advance network functions such as provision-

M. Tsujino (✉)
NTT Network Technology Laboratories, 3-9-11 Midori-Cho,
Musashino-Shi, Tokyo 180-8585, Japan
e-mail: masayuki.tsujino.ph@hco.ntt.co.jp

R. Lee (ed.), *Big Data, Cloud Computing, and Data Science Engineering*,
Studies in Computational Intelligence 844,
https://doi.org/10.1007/978-3-030-24405-7_1

ing delay-sensitive services. In 5G, improvement in network efficiency is expected by assigning radio control units to fog computers under the Centralized Radio Access Network (C-RAN) architecture [32]. For this reason, telecom carriers need to study an effective model of designing an IoT infrastructure from a strategic perspective while considering future service trends.

I focused on designing the placement and capacity of both cloud and fog computers and the communication links to support this design. The placement of the computing function affects the traffic flow on a communication network; therefore, an integrated model for designing both computers and communication links should be developed. In addition, the progress in increasing the speed of a wide area network is slower compared with that of CPU processing [20]. Therefore, it is more important to consider integrated design in the future. Furthermore, we should efficiently assign workloads to cloud/fog computers by considering their respective features; a cloud computer can offer scalable processing services, while a fog computer meets the requirement for delay-sensitive services [11, 26]. Therefore, I focus on developing a model for designing IoT infrastructures consisting of cloud, fog, and communication layers under this combined cloud-fog paradigm.

The design of this infrastructure is generally based on predicted future demands. However, it is extremely difficult to predict the future demands of innovative IoT services. For example, the Japanese Information Communication White Paper (2017 edition) [14] states that the base and economic growth scenarios regarding the impact of IoT and AI on the GDP of Japan are expected to deviate by about 22% in 2030 (725 trillion yen/593 trillion yen). Also, it is much more difficult to predict the demand at each regional site, which is used as basic data for placement/capacity design. This is because the generation of demand is often not derived from human activities in IoT services. Moreover, there are many different use cases in various industries, unlike consumer-targeted services. Therefore, we need to develop a robust design model against demand uncertainty.

Thus, I propose a robust model for designing IoT infrastructures assuming the worst demand situation. This model is formulated as a robust optimization problem. Since the concept of robust optimization proposed by Ben-Tal and Nemirovski [4], it has been extensively studied in both theory and application [5, 25, 29, 33]. Therefore, I also present a method of solving the robust optimization problem formulated from the proposed model, which is practically difficult to solve.

The contributions of this paper are as follows.

- I propose a robust design model formulated as a robust optimization problem for the placement/capacity design of IoT infrastructures consisting of cloud, fog, and communication layers under uncertain demands.
- I present a method of solving this robust optimization problem by translating it into a deterministic equivalent optimization problem called “robust counterpart.”
- I show the effectiveness of applying the proposed robust design model to a real network, its effectiveness in enhancing robustness against demand uncertainty, and its scalability for large-scale networks from the viewpoint of computing time.

This paper is organized as follows. After discussing related work in Sect. 2, I formulate the proposed robust design model as a robust optimization problem in Sect. 3. In Sect. 4, I present the method of solving this robust optimization problem by translating it into a deterministic equivalent optimization problem, which is the robust counterpart to this problem. In Sect. 5, I discuss the evaluation results from numerical experiments conducted to validate my model and method. Finally, I conclude the paper and briefly comment on future work in Sect. 6.

2 Related Work

Various studies have been conducted on the problem of capacity management. The appropriate method of assigning workloads and traffic to computers and communication links are chosen under the condition that the placement and capacity are fixed.

The assignment of traffic to communication links is called "traffic engineering (TE)," and extensive research has been conducted on this. The route-computation algorithm, which is the most important topic of TE in theory, is described in detail in a previous paper [24]. Several studies [8, 15, 30] investigated robust TE technologies against uncertain demands through various approaches. Accommodating as many virtual network embeddings (VNEs) as possible based on the robust optimization approach on the premise that the demand for constructing VNEs is uncertain has been studied [9].

Various studies focused on capacity management targeting the assignment to computing functions as well as communication links. Several methods have been proposed for effectively sharing workloads among fog nodes based on the usage status of fog computers and communication links to assign more workloads [1, 21, 22]. In consideration of the three-layer structure under the combined cloud-fog paradigm, methods have been proposed for determining which cloud/fog computer to which a workload should be assigned from the request level and load condition [28]. Also, a method for stably controlling the assigning of content-access requests to cache servers and communication links against demand fluctuation is being studied within the context of content delivery networks (CDNs) [16].

Studies on the network-design problem differ from those on the capacity-management problem. The network-design problem is focused on determining the desired placement and capacity based on the given conditions of demand and cost. The problem of designing communication networks is described as "topological design" [24]. This network-design problem is NP-complete, as discussed in a previous paper [17], which is a pioneering theoretical work. Therefore, it is generally considered that the network-design problem is more difficult than the capacity-management problem. Regarding the design problem for communication networks, a study investigating a method of determining the placement of communication links with the robust optimization approach under uncertain demands between sites has been conducted [3]. However, this study did not focus on determining the appropriate capacity. There has been little research on designing both computing functions and communication links, even on the premise of the determined demand [27].

I address the problem of designing the placement and capacity of IoT infrastructures consisting of cloud, fog, and communication layers under uncertain demand, which has not been examined and analyzed.

3 Problem Formulation

I present the formulation of the proposed robust design model as a robust optimization problem of IoT infrastructures in this section. The proposed model introduces the placement and capacity of cloud/fog computers and communication links, which can be constructed with minimum cost to satisfy any demand estimated as uncertainty.

3.1 System Architecture of IoT Infrastructure

The system architecture of the target IoT infrastructure for this study is shown in Fig. 1.

The IoT devices connect to the communication network via the IoT gateway for relaxing the load of the IoT infrastructure and seamlessly integrating individual IoT environments. The demands for assigning workloads to cloud/fog computers (cloud/fog demands) are aggregated by the IoT gateway.

In the combined cloud-fog paradigm, the cloud/fog demand aggregated by the IoT gateway is assigned to the second fog layer or the third cloud layer through the first communication layer. I assume that it is possible to configure the cloud/fog computers and communication links to be assigned under the management architecture of an SDN.

3.2 Input Parameters

Configuration of IoT infrastructure

The network topology of the IoT infrastructure is represented by a (directed) graph $G = \{N, A\}$, where N is the set of nodes on the network, and A is the set of (directed) links. The nodes include the sites where the cloud/fog computer can be placed (cloud/fog node) and/or those where the IoT gateway and/or routing devices are placed (several types of equipment can be placed on each site). Also, the links indicate that transmission equipment can be installed between both their end nodes. I denote the set of sites where the IoT gateways can be placed, which are the sources

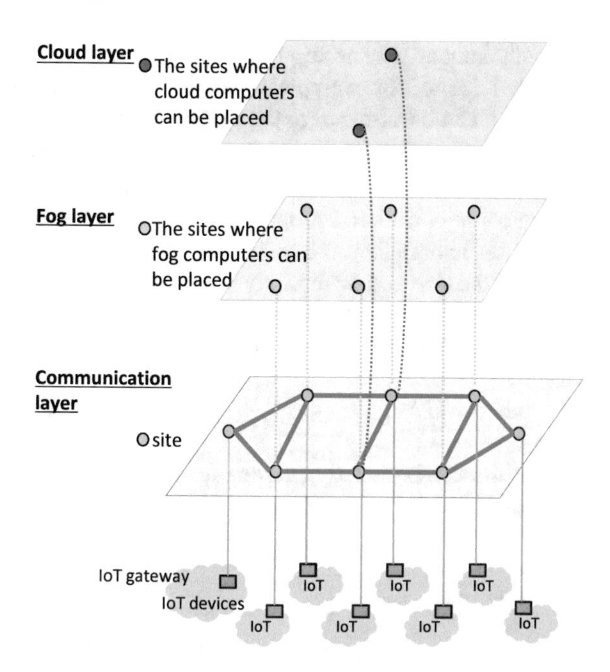

Fig. 1 System architecture of target IoT infrastructure

of cloud/fog demands, as N_{gw}. Also, the sets of the cloud and fog nodes are denoted as N_{cl} and N_{fg}, respectively.

As mentioned above, I take into account the flexible assignment of the demands between cloud and fog computers. I next describe my model by exemplifying a case in which a fog computer can be assigned to cloud demand, which is practically meaningful.

Conditions associated with design costs

In addition to the variable costs imposed depending on the capacity of traffic or workload, I consider the fixed design costs associated with the placement of communication links or cloud/fog computers. The notation $b_{lk}^{(a)}$ is the fixed design cost for placing transmission equipment on the communication link $a \in A$, and $c_{lk}^{(a)}$ is the variable cost coefficient per unit traffic. Similarly, the fixed design cost for placing cloud/fog computers at a cloud/fog node $n \in N_{cl}/N_{fg}$ is denoted as $b_{cl}^{(n)}/b_{fg}^{(n)}$, and the variable cost coefficient per unit workload is denoted as $c_{cl}^{(n)}/c_{fg}^{(n)}$.

Conditions for uncertain demands

The set of cloud/fog demands over the entire network is K_{cl}/K_{fg}, and the total set of these disjoint demand sets is $K = K_{cl} \cup K_{fg} (disjoint)$. Let $gw^{(k)}$ be the source node for each demand k. The fog computers that can be assigned for each demand are restricted due to, for example, delay. To represent these restrictions, the set of fog nodes that can be assigned to each k is denoted as $N_{fg}^{(k)} \subset N_{fg}$.

Let $d^{(k)}$ (treated as a variable in the model) be the amount of traffic demanded on communication links for each k. To simplify the model, the workload at cloud/fog computers is considered proportional to the traffic of each demand ($d^{(k)}$), and its proportional coefficient is expressed as $h^{(k)}$.

I treat the demand uncertainty with the uncertain demand set D specified by the parameter of robustness, $\Gamma \geq 0$, which was proposed by Bertismas and Nemirovski [6], as shown below.

$$D := \{d^{(k)} \mid \bar{d}^{(k)} \leq d^{(k)} \leq \bar{d}^{(k)} + \tilde{d}^{(k)}, k \in K; \textstyle\sum_{k \in K} \frac{d^{(k)} - \bar{d}^{(k)}}{\tilde{d}^{(k)}} \leq \Gamma\}, \quad (1)$$

The above-mentioned notations for the input parameters in my model are listed in Table 1.

Table 1 Input parameters

Parameter	Description
$G = \{N, A\}$	Target communication network
N	Set of network nodes
N_{gw}	Set of network nodes where IoT gateways are placed, $N_{gw} \subset N$
N_{cl}	Set of network nodes where cloud computers can be placed, $N_{cl} \subset N$
N_{fg}	Set of network nodes where fog computers can be placed, $N_{fg} \subset N$
A	Set of network links
$b_{lk}^{(a)}$	Fixed design cost at link $a \in A$
$c_{lk}^{(a)}$	Traffic-dependent unit cost at link $a \in A$
$b_{cl}^{(n)}$	Fixed design cost for placing cloud computers at node $n \in N_{cl}$
$c_{cl}^{(n)}$	Workload-dependent unit cost for assigning cloud computers at node $n \in N_{cl}$
$b_{fg}^{(n)}$	Fixed design cost for placing fog computers at node $n \in N_{fg}$
$c_{fg}^{(n)}$	Workload-dependent unit cost for assigning fog computers at node $n \in N_{fg}$
K_{cl}	Set of demands for assigning cloud computers
K_{fg}	Set of demands for assigning fog computers
K	Set of all demands, $K = K_{cl} \cup K_{fg}$
$gw^{(k)}$	Source node for demand k
$N_{fg}^{(k)}$	Set of nodes that can be assigned for each demand of fog computers k, $N_{fg}^{(k)} \subset N_{fg}$
$h^{(k)}$	Proportional factor of workload for demand $k \in K$
$\bar{d}^{(k)}$	Average traffic for demand $k \in K$
$\tilde{d}^{(k)}$	Maximum deviation from average traffic for demand $k \in K$
Γ	Robustness-level parameter

Table 2 Variables

Variable	Description
$x_{lk}^{(a)}$	Capacity of transmission-link equipment installed at link $a \in A$
$x_{cl}^{(n)}$	Capacity of cloud computers installed at cloud node $n \in N_{cl}$
$x_{fg}^{(n)}$	Capacity of fog computers installed at fog node $n \in N_{fg}$
$y_{lk}^{(k,a)}$	Whether demand $k \in K$ is assigned to link $a \in A$
$y_{cl}^{(k,n)}$	Whether demand $k \in K_{cl}$ is assigned to cloud or fog node $n \in N_{cl} \cup N_{fg}$
$y_{fg}^{(k,n)}$	Whether demand $k \in K_{fg}$ is assigned to fog node $n \in N_{fg}^{(k)}$
$z_{lk}^{(a)}$	Whether link $a \in A$ is placed
$z_{cl}^{(n)}$	Whether cloud computers are installed at cloud node $n \in N_{cl}$
$z_{fg}^{(n)}$	Whether fog computers are installed at fog node $n \in N_{fg}$
$d^{(k)}$	Volume for demand $k \in K$

3.3 Variables

I now describe the variables in the proposed robust design model. As the control variables to be targeted in the decision making of the design model, I describe the capacity of the equipment to be installed as "x", the choice of whether to assign the demand to equipment as the binary variable represented as "y", the choice of whether to place equipment as the binary variable represented as "z", and $d^{(k)}$, which represents the uncertain demand value previously explained, is also treated as a variable in the model.

Specifically, these variables are listed in Table 2.

In these variables, I use vector notations such as $\boldsymbol{x} = \{x_{lk}^{(a)}, x_{cl}^{(n_{cl})}, x_{fg}^{(n_{fg})} \mid a \in A, n_{cl} \in N_{cl}, n_{fg} \in N_{fg}\}$.

3.4 Robust Design Model

The robust design model is formulated as follows:

Objective function

The total design cost for constructing IoT infrastructures consists of the fixed design cost and variable cost corresponding to the capacity of the communication link and cloud/fog computer to be installed. Thus, the objective function of the model is

$$\begin{aligned} F(\boldsymbol{x}, z) =& \boldsymbol{b} \cdot z + \boldsymbol{c} \cdot \boldsymbol{x} \\ =& \sum_{a \in A} b_{lk}^{(a)} \cdot z_{lk}^{(a)} + \sum_{n \in N_{cl}} b_{cl}^{(n)} \cdot z_{cl}^{(n)} + \sum_{n \in N_{fg}} b_{fg}^{(n)} \cdot z_{fg}^{(n)} \\ &+ \sum_{a \in A} c_{lk}^{(a)} \cdot x_{lk}^{(a)} + \sum_{n \in N_{cl}} c_{cl}^{(n)} \cdot x_{cl}^{(n)} + \sum_{n \in N_{fg}} c_{fg}^{(n)} \cdot x_{fg}^{(n)} \end{aligned} \tag{2}$$

Conditions for satisfying demands

These formulas represent that each cloud/fog demand can be assigned to a computer, which is installed at assignable sites. The second term on the left side of (3a) follows the premise that a fog computer can be assigned to a cloud demand.

$$\sum_{n \in N_{cl}} y_{cl}^{(k,n)} + \sum_{n \in N_{fg}} y_{fg}^{(k,n)} = 1, for\ k \in K_{cl} \tag{3a}$$

$$\sum_{n \in N_{fg}^{(k)}} y_{fg}^{(k,n)} = 1, for\ k \in K_{fg} \tag{3b}$$

Flow conservation rules

These formulas represent the flow conservation rules on the communication network for cloud/fog demands. Note that (4b) follows the premise that a fog computer can be assigned to a cloud demand.

$$\sum_{a \in out(n)} y_{lk}^{(k,a)} - \sum_{a \in in(n)} y_{lk}^{(k,a)} = \begin{cases} -y_{cl}^{(k,n)}\ if\ n \in N_{cl} & (4a) \\ -y_{fg}^{(k,n)}\ if\ n \in N_{fg} & (4b) \\ 1 \quad if\ n = gw^{(k)} & (4c) \\ 0 \quad otherwise, & (4d) \end{cases}$$

$$for\ n \in N, k \in K_{cl}$$

$$\sum_{a \in out(n)} y_{lk}^{(k,a)} - \sum_{a \in in(n)} y_{lk}^{(k,a)} = \begin{cases} -y_{fg}^{(k,n)}\ if\ n \in N_{fg}^{(k)} & (5a) \\ 1 \quad if\ n = gw^{(k)} & (5b) \\ 0 \quad otherwise, & (5c) \end{cases}$$

$$for\ n \in N, k \in K_{fg},$$

where $in(n)/out(n)$ represents the set of links entering/exiting $n \in N$, respectively.

Capacity constraints

Traffic flowing through the communication link and workload assigned to the cloud/fog node are restricted by the capacity of the link and cloud/fog node.

$$\sum_{k \in K} d^{(k)} \cdot y_{lk}^{(k,a)} \leq x_{lk}^{(a)}, for\ a \in A \tag{6a}$$

$$\sum_{k \in K_{cl}} h^{(k)} \cdot d^{(k)} \cdot y_{cl}^{(k,n)} \leq x_{cl}^{(n)}, for\ n \in N_{cl} \tag{6b}$$

$$\sum_{k \in K} h^{(k)} \cdot d^{(k)} \cdot y_{fg}^{(k,n)} \leq x_{fg}^{(n)}, for\ n \in N_{fg} \tag{6c}$$

Placement constraints

When some traffic/workloads are assigned to the link/node, the equipment should be placed. These formulas represent the constraints for placement. They are based on the big-M method [12], and it is necessary to set parameter M to a sufficiently large value so that the constraint expression always holds when $z = 1$.

$$x_{lk}^{(a)} \leq M \cdot z_{lk}^{(a)}, for\ a \in A \tag{7a}$$

$$x_{cl}^{(n)} \leq M \cdot z_{cl}^{(n)}, for\ n \in N_{cl} \tag{7b}$$

$$x_{fg}^{(n)} \leq M \cdot z_{fg}^{(n)}, for\ n \in N_{fg} \tag{7c}$$

Control variable definitions

This has the following constraints from the definition of the control variable.

$$\boldsymbol{x} \geq 0,\ \boldsymbol{y}, \boldsymbol{z} \in \{0, 1\}. \tag{8}$$

With respect to $\boldsymbol{x}, \boldsymbol{y}, \boldsymbol{z}$, I denote a constraint set derived from constraints (3)–(5), (7), and (8), which are independent of $\boldsymbol{d}$, as S, and a constraint set derived from constraint (6), which is dependent on $\boldsymbol{d}$, as $T(\boldsymbol{d})$.

Thus, my robust design model of IoT infrastructures can be formulated as the following robust optimization problem for finding the minimum cost guaranteed for any demand pattern included in the uncertain demand set D.

$$\min : F(\boldsymbol{x}, \boldsymbol{z}) \tag{9}$$

$$\text{s.t.} : \{\boldsymbol{x}, \boldsymbol{y}, \boldsymbol{z}\} \in S \cap T(\boldsymbol{d}),\ \forall \boldsymbol{d} \in D \tag{10}$$

4 Robust Counterpart for Proposed Model

The proposed robust design model is different from the ordinary optimization problem. Therefore, it is practically difficult to solve the model in this form. In this section, I derive a "robust counterpart" of the model based on the approach proposed in a previous paper [6]. The robust counterpart is a deterministic optimization problem equivalent to the robust optimization problem formulated from the proposed model.

First, consider the demand that maximizes the objective function. From constraints (1) and (6a), the capacity of communication link $a \in A$, which increases the objective function, is considered as the following problem $P_{lk}^{(a)}$, $a \in A$;

$$\max : \sum_{k \in K} \tilde{d}^{(k)} \cdot y_{lk}^{(k,a)} \cdot w^k \tag{11}$$

$$\text{s.t.} : \sum_{k \in K} w^{(k)} \leq \Gamma \; [\pi_{lk}^{(a)}] \tag{12}$$

$$w^{(k)} \leq 1, for\ k \in K \; [\rho_{lk}^{(k,a)}] \tag{13}$$

$$w^{(k)} \geq 0, for\ k \in K, \tag{14}$$

where I convert variables to be used in the following to simplify the model description.

$$w^{(k)} \rightarrow \frac{d^{(k)} - \bar{d}^{(k)}}{\tilde{d}^{(k)}} \tag{15}$$

The following dual problem $Q_{lk}^{(a)}$ of $P_{lk}^{(a)}$ is obtained with the dual variables $\pi^{(a)}$ and $\rho^{(k,a)}$ for constraints (12) and (13), respectively.

$$\min : \Gamma \cdot \pi_{lk}^{(a)} + \sum_{k \in K} \rho_{lk}^{(k,a)} \tag{16}$$

$$\text{s.t.} : \pi_{lk}^{(a)} + \rho_{lk}^{(k,a)} \geq \tilde{d}^{(k)} \cdot y_{lk}^{(k,a)}, for\ k \in K \tag{17}$$

$$\pi_{lk}^{(a)} \geq 0 \tag{18}$$

$$\rho_{lk}^{(k,a)} \geq 0, for\ k \in K \tag{19}$$

The dual problems for cloud node $n \in N_{cl}$ and fog node $n \in N_{fg}$ are obtained in the same manner.

The robust counterpart, which is equivalent to the robust optimization problem formulated from the proposed robust design model, can be derived with these dual problems as follows:

$$\min : \boldsymbol{b} \cdot \boldsymbol{z} \tag{20}$$
$$+ \sum_{a \in A} c_{lk}^{(a)} \cdot \{ \sum_{k \in K} \bar{d}^{(k)} \cdot y_{lk}^{(k,a)} + \Gamma \cdot \pi_{lk}^{(a)} + \sum_{k \in K} \rho_{lk}^{(k,a)} \}$$

$$+\sum_{n\in N_{cl}} c_{cl}^{(n)} \cdot \{\sum_{k\in K_{cl}} \bar{d}^{(k)} \cdot y_{cl}^{(k,n)} + \Gamma \cdot \pi_{cl}^{(n)} + \sum_{k\in K_{cl}} \rho_{cl}^{(k,n)}\}$$

$$+\sum_{n\in N_{fg}} c_{fg}^{(n)} \cdot \{\sum_{k\in K} \bar{d}^{(k)} \cdot y_{fg}^{(k,n)} + \Gamma \cdot \pi_{fg}^{(n)} + \sum_{k\in K} \rho_{fg}^{(k,n)}\}$$

$$\text{s.t.} :\{\boldsymbol{x}, \boldsymbol{y}, \boldsymbol{z}\} \in S$$

$$\pi_{lk}^{(a)} + \rho_{lk}^{(k,a)} \geq \tilde{d}^{(k)} \cdot y_{lk}^{(k,a)}, for\ k \in K \tag{21}$$

$$\pi_{cl}^{(n)} + \rho_{cl}^{(k,n)} \geq h^{(k)} \cdot \tilde{d}^{(k)} \cdot y_{cl}^{(k,n)}, for\ k \in K_{cl} \tag{22}$$

$$\pi_{fg}^{(n)} + \rho_{fg}^{(k,n)} \geq h^{(k)} \cdot \tilde{d}^{(k)} \cdot y_{fg}^{(k,n)}, for\ k \in K \tag{23}$$

$$\boldsymbol{\pi}, \boldsymbol{\rho} \geq 0 \tag{24}$$

Since this problem is a mixed integer linear programming problem, which is tractable in computation, it is possible to solve with a general-purpose optimization solver.

5 Evaluation

In this section, I explain the evaluation results from numerical experiments I conducted by applying the proposed model to a real network model (JPN 48) and mathematical network model (Watts-Strogatz (WS)). The following conditions were common for both models.

- Only one node of installing cloud computers was given for each model (placement of cloud computers was not included in the design).
- The groups of fog nodes composed of adjacent nodes ("fog group") was given for each model, and the workload for fog demand was assigned within this group.
- Demands for assigning workload to both cloud and fog computers were generated at all nodes.
- Fog demand was determined as a uniform random number so that it would be on average 1/2 the cloud demand at each node.
- The workload was assumed to be the same as the traffic generated for each demand ($h^{(k)} = 1, k \in K$).
- The maximum deviation from average traffic/workload for each demand ($\tilde{d}^{(k)}, k \in K$) was determined with a uniform random number so that it would become 1/3 the average traffic ($\bar{d}^{(k)}, k \in K$) on average.
- The ratio between the fixed design cost and traffic/workload-dependent unit cost ($b_{lk}^{(a)}/c_{lk}^{(a)}, a \in A, b_{cl}^{(n)}/c_{cl}^{(n)}, n \in N_{cl}, b_{fg}^{(n)}/c_{fg}^{(n)}, n \in N_{fg}$, referred to as the "fixed-variable (f–v) ratio"), was fixed regardless of the cloud/fog nodes and communication links for each model.

Table 3 Fog group in JPN 48

Fog group Name	Node (prefecture)
1. Hokkaido-Tohoku	Hokkaido, Aomori, Iwate, Miyagi, Akita, Yamagata, Fukushima
2. Kanto	Ibaraki, Tochigi, Gunma, Saitama, Chiba, Tokyo (east), Tokyo (west), Kanagawa, Yamanashi
3. Chubu	Niigata, Toyama, Ishikawa, Fukui, Nagano, Gifu, Shizuoka, Aichi
4. Kinki	Mie, Shiga, Kyoto, Osaka, Hyogo, Nara, Wakayama
5. Chugoku-Shikoku	Tottori, Shimane, Okayama, Hiroshima, Yamaguchi, Tokushima, Kagawa, Ehime, Kochi
6. Kyushu	Fukuoka, Saga, Nagasaki, Kumamoto, Oita, Miyazaki, Kagoshima, Okinawa

5.1 Case Study 1 JPN 48 Model

I evaluated the proposed model with JPN 48 [2], which was created by assuming Japan's national communication network for the purpose of evaluating the network design/control method.

The cloud node was placed in Tokyo (east), i.e., the central wards of Tokyo. I constructed the fog group shown in Table 3 based on the conventional Japanese regional divisions. I set the link distance indicated with JPN 48 as the variable cost coefficient on communication links, 500 as the variable cost coefficient of cloud nodes and 250 as the variable cost coefficient of fog nodes. The f–v ratio was changed according to each evaluation. In addition, the average value of cloud demand from each node was determined from "population under node" indicated with JPN 48.

Figure 2 shows the capacity of the communication link (left figure), that of the fog node for fog demand (middle figure), and that of the node belonging to each fog group for cloud demand (right figure) when the Γ of the robustness level parameter was set to five (5), where fog group #7 became a cloud node. Note that zero (0) for the capacity means that the fog computer was not placed, and no equipment was installed.

I found that the number of sites where fog computers were placed decreased as the f–v ratio increased. This is because the workload for cloud/fog demand was concentrated on specific nodes to reduce the number of sites for fog computers, even though it led to an increase in communication traffic (left figure). I can assume that the model may be able to exhibit characteristics of a real design problem from one example.

Next, I evaluated demand satisfiability through Monte Carlo simulation on the proposed robust design model with an f–v ratio of 500 and changing Γ in the range of 0–5. I generated $d^{(k)}, k \in K$ for each cloud/fog demand as a uniform random number in the range of $[\bar{d}^{(k)} - \tilde{d}^{(k)}, \bar{d}^{(k)} + \tilde{d}^{(k)}]$ and assigned it to a cloud/fog node and communication link. I conducted 1,000 trials for each robust design specified by Γ.

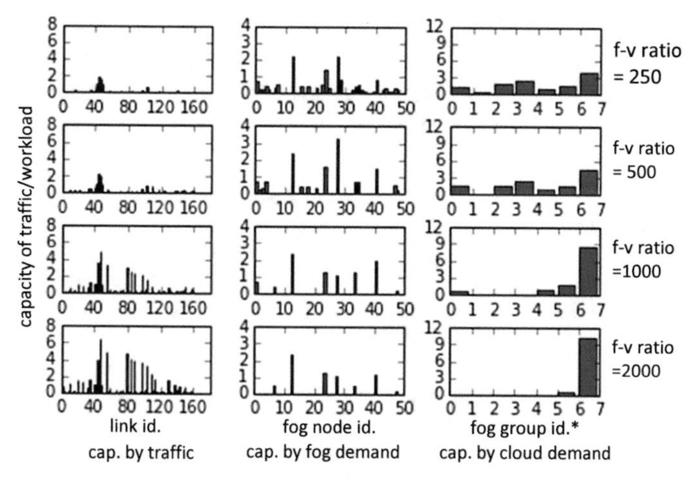

Fig. 2 Capacities of cloud, fog, and communication layers designed with JPN 48

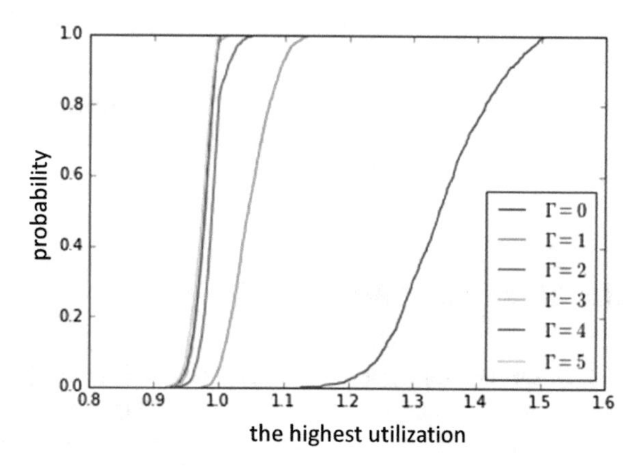

Fig. 3 Evaluation of demand satisfiability for robust design

Figure 3 shows the distribution of the highest utilization rate among the cloud/fog nodes and communication links for each trial. Table 4 shows the "cost of robustness (CoR) ratio" for each robust design. The CoR ratio is defined as the ratio of the robust design cost divided by the cost of $\Gamma = 0$ ("non-robust design") and indicates the cost required for gaining robustness. These results indicate that it is possible to prevent the capacity from being exceeded when increasing Γ while sacrificing cost.

I also adjusted the capacities for the design of $\Gamma = 1, 2, 3, 4$ to match the design costs among the compared designs. The capacities were expanded for each design to make the design cost equal to that of $\Gamma = 5$, while maintaining the ratio of capacity among communication links and computers. Evaluation on demand satisfaction

Table 4 Cost of robustness ratio

$\Gamma = 0$	$\Gamma = 1$	$\Gamma = 2$	$\Gamma = 3$	$\Gamma = 4$	$\Gamma = 5$
1.000	1.099	1.159	1.196	1.214	1.228

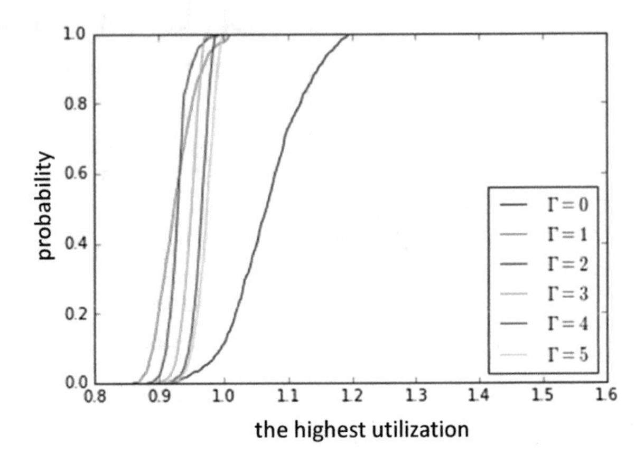

Fig. 4 Evaluation of cost efficiency for proposed robust design model

through Monte Carlo simulation was also conducted for this design, and the results are shown in Fig. 4. These simulations demonstrated the high cost performance of the proposed robust design model.

5.2 *Case Study 2-WS Model*

I used the network generated using the WS model [31] to evaluate the proposed model in various networks with different shapes and scales. Among the parameters specifying the WS model, "average degree" and "rewiring rate" were fixed to 4 and 0.1, respectively, and "number of nodes" was treated as a scale parameter indicating the scale of the problem. A site where cloud computers can be placed was randomly chosen for each case. The fog group was composed of five adjacent nodes. The average value of cloud demand was generated from a random number based on a normal distribution with an average of 200 and standard deviation of 60. The variable cost factor was determined by a uniform random number, and the f–v ratio was set to 100.

By changing the number of nodes, or scale, in the range of 20– 180, I introduced 10 cases to be evaluated for each scale. Robust designs in the Γ range of 0–5 were obtained.

In the same manner as the evaluation illustrated in Fig. 4, the capacities were expanded for each design of $\Gamma = 1, 2, 3, 4$ to make the design cost equal to that of $\Gamma = 5$. Figure 5 shows the average (median) and standard deviation (error bar) in the number of trials, at which capacity was exceeded in any of the 1,000 trials,

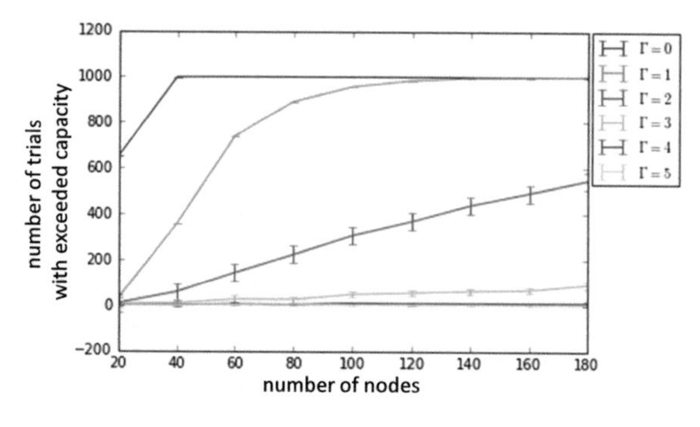

Fig. 5 Evaluation of cost efficiency for robust design under various conditions

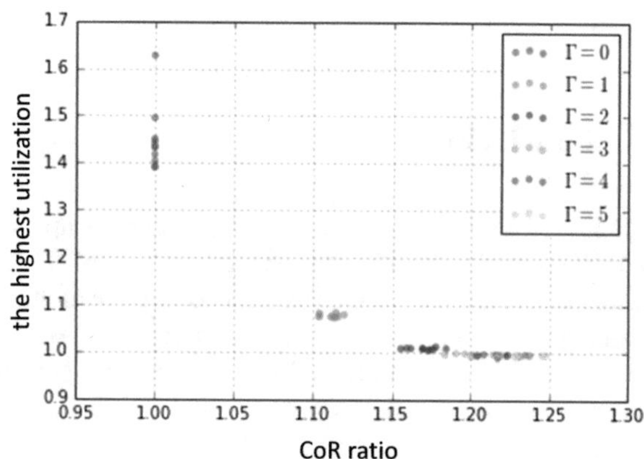

Fig. 6 Relationship between CoR ratio and demand satisfiability

among 10 cases for each scale by Monte Carlo simulation against this modified design. Although it is the result of simulation based on the same design cost, as Γ increased, the rate of causing excess capacity decreased. Therefore, the proposed robust design model is effective under various network conditions because it leads to a robust design with high cost performance.

In Fig. 6, the CoR ratio (x axis) and the highest utilization rate (y axis) are plotted on a scatter gram for the 10 cases in which the number of nodes was 100 when designing an IoT infrastructure by changing Γ from 0 to 5 and evaluated through 1,000 trials of Monte Carlo simulation. I confirmed the relation between the cost of robustness and demand satisfiability.

Finally, Fig. 7 shows the computation time associated with the number of nodes. The computation was carried out with an Intel Core i7-4790K CPU @ 4.00 GHz and general purpose optimization solver "Gurobi 8.1" [13]. The average computation times are indicated on the y axis for each scale when designing the 10 cases with $\Gamma = 5$. Therefore, the proposed model can be applied to large-scale problems of about 200 nodes using the latest optimization solver.

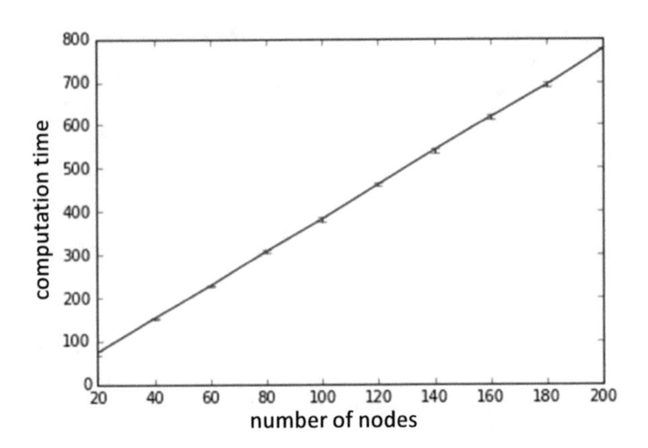

Fig. 7 Computation time

6 Conclusion

I proposed a robust design model for placement/capacity of IoT infrastructures under uncertain demands. The system architecture of the target IoT infrastructure consists of three layers; cloud, fog, and communication, under the combined cloud-fog paradigm. The proposed model was formulated as a robust optimization problem, and I presented a method of solving this problem. My numerical experiments indicate the effectiveness of the proposed model and the possibility of applying the method to the proposed model to a practical scaled network.

Further studies include developing methods for applying diversified demand conditions and evaluating the proposed robust design model under conditions of actual IoT infrastructure design.

References

1. Abedin, S.F., Alam, M.G.R., Tran, N.H., Hong, C.S.: A fog based system model for cooperative IoT node pairing using matching theory. In: The 17th Asia-Pacific Network Operations and Management Symposium (APNOMS)
2. Arakawa, S., Sakano, T., Tukishima, Y., Hasegawa, H., Tsuritani, T., Hirota, Y., Tode, H.: Topological characteristic of Japan photonic network model. IEICE Tech. Rep. **113**(91), 7–12 (2013) (in Japanese)
3. Bauschert, T., Bsing, C., D'Andreagiovanni, F., Koster, A.C.A., Kutschka, M., Steglich, U.: Network planning under demand uncertainty with robust optimization. IEEE Commun. Mag. **52**(2), 178–185 (2014)
4. Ben-tal, A., Nemirovski, A.: Robust solutions of linear programming problems contaminated with uncertain data. Math. Progr. **88**, 411–424 (2000)
5. Ben-tal, A., Ghaoui, L.E., Nemirovski, A.: Robust Optimization. Princeton Series in Applied Mathematics. Princeton University Press, Princeton (2009)
6. Bertsimas, D., Sim, M.: The price of robustness. Oper. Res. **52**(1), 35–53 (2004)

7. Bonomi, F., Milito, R., Zhu, J., Addepalli, S.: Fog computing and its role in the internet of things. In: Proceedings of the First Edition of the MCC Workshop on Mobile Cloud Computing, pp. 13–16. ACM (2012)
8. Chandra, B., Takahashi, S., Oki, E.: Network congestion minimization models based on robust optimization. IEICE Trans. Commun. **E101.B**(3), 772–784 (2018)
9. Coniglio, S., Koster, A., Tieves, M.: Data uncertainty in virtual network embedding: robust optimization and protection levels. J. Netw. Syst. Manag. **24**(3), 681–710 (2016)
10. Evans, D.: The internet of things: how the next evolution of the internet is changing everything. CISCO White Paper (2011)
11. Ghosh, R., Simmhan, Y.: Distributed scheduling of event analytics across edge and cloud. ACM TCPS | ACM Trans. Cyberphysical Syst. **2**(4), 1–28 (2018)
12. Griva, I., Nash, S.G., Sofer, A.: Linear and Nonlinear Optimization. Society for Industrial and Applied Mathematics, 2nd edn. (2009)
13. http://www.gurobi.com
14. Information and Communication in Japan. Ministry of Internal Affairs and Communication, Japan (2017)
15. Jain, S., Kumar, A., Mandal, S., Ong, J., Poutievski, L., Singh, A., Venkata, S., Wanderer, J., Zhou, J., Zhu, M., Zolla, J., Hlzle, U., Stuart, S., Vahdat, A.: B4: experience with a globally-deployed software defined WAN. ACM Spec. Interes. Group Data Commun. (SIGCOMM) **2013**, 3–14 (2013)
16. Kamiyama, N., Takahashi, Y., Ishibashi, K., Shiomoto, K., Otoshi, T., Ohsita, Y., Murata, M.: Optimizing cache location and route on CDN using model predictive control. In: The 27th International Teletraffic Congress (ITC), pp. 37–45 (2015)
17. Magnanti, T.L., Wong, R.T.: Network design and transportation planning: models and algorithms. Transp. Sci. **18**(1), 1–55 (1984)
18. Mukherjee, M., Shu, L., Member, S., Wang, D.: Survey of fog computing: fundamental, network applications, and research challenges. IEEE Commun. Surv. Tutor. **20**(3), 1826–1857 (2018)
19. Mouradian, C., Naboulsi, D., Yangui, S., Glitho, R.H., Morrow, M.J., Polakos, P.A.: A comprehensive survey on fog computing: state-of-the-art and research challenges. IEEE Commun. Surv. Tutor. **20**(1), 416–464 (2018)
20. Nielsen's law of internet bandwidth. http://www.nngroup.com/articles/law-of-bandwidth/
21. Nishio, T., Shinkuma, R., Takahashi, T., Mandayam, N.B.: Service-oriented heterogeneous resource sharing for optimizing service latency in mobile cloud. In: Proceedings of the First International Workshop on Mobile Cloud Computing & Networking (MobileCloud '13), pp. 19–26 (2013)
22. Oueis, J., Strinati, E.C., Sardellitti, S., Barbarossa, S.: Small cell clustering for efficient distributed fog computing: a multi-user case. In: The 82nd Vehicular Technology Conference (VTC2015-Fall), pp. 1–5 (2015)
23. Perera, C., Harold, C., Member, L., Jayawardena, S.: The emerging internet of things marketplace from an industrial perspective: a survey. IEEE Trans. Emerg. Top. Comput. **3**(4), 585–598
24. Pióro, M., Medhi, D.: Routing, Flow, and Capacity Design in Communication and Computer Networks. Morgan Kaufmann, San Francisco (2004)
25. Shabanzadeh, M., Sheikh-El-Eslami, M.K., Haghifam, M.R.: The design of a risk-hedging tool for virtual power plants via robust optimization approach. Appl. Energy **155**, 766–777 (2015)
26. Souza, V.B.C., Ramrez, W., Masip-Bruin, X., Marn-Tordera, E., Ren, G., Tashakor, G.: Handling service allocation in combined fog-cloud scenarios. In: 2016 IEEE International Conference on Communications (ICC), pp. 1–5 (2016)
27. Takeshita, K., Shiozu, H., Tsujino, M., Hasegawa, H.: An optimal server-allocation method with network design problem. In: Proceedings of the 2010 IEICE Society Conference, vol. 2010, issue 2, p. 93 (2010) (in Japanese)
28. Taneja, M., Davy, A.: Resource aware placement of IoT application modules in fog-cloud computing paradigm. In: 2017 IFIP/IEEE Symposium on Integrated Network and Service Management (IM), pp. 1222–1228 (2017)

29. Ttnc, R.H., Koenig, M.: Robust asset allocation. Ann. Oper. Res. **132**(1–4), 157–187 (2000)
30. Wang, H., Xie, H., Qiu, L., Yang, Y.R., Zhang, Y., Greenberg, A.: COPE: traffic engineering in dynamic networks. ACM Spec. Interes. Group Data Commun. (SIGCOMM) **2006**, 99–110 (2006)
31. Watts, D.J., Strogatz, S.H.: Collective dynamics of "small-world" networks. Nature **393**, 440–442 (1998)
32. Yang, P., Zhang, N., Bi, Y., Yu, L., Shen, X.S.: Catalyzing cloud-fog interoperation in 5G wireless networks: an SDN approach. IEEE Netw. **31**(5), 14–21 (2017)
33. Yu, C.S., Li, H.L.: A robust optimization model for stochastic logistic problems. Int. J. Prod. Econ. **64**(1–3), 385–397 (2000)

Multi-task Deep Reinforcement Learning with Evolutionary Algorithm and Policy Gradients Method in 3D Control Tasks

Shota Imai, Yuichi Sei, Yasuyuki Tahara, Ryohei Orihara and Akihiko Ohsuga

Abstract In deep reinforcement learning, it is difficult to converge when the exploration is insufficient or a reward is sparse. Besides, on specific tasks, the amount of exploration may be limited. Therefore, it is considered effective to learn on source tasks that were previously for promoting learning on the target tasks. Existing researches have proposed pretraining methods for learning parameters that enable fast learning on multiple tasks. However, these methods are still limited by several problems, such as sparse reward, deviation of samples, dependence on initial parameters. In this research, we propose a pretraining method to train a model that can work well on variety of target tasks and solve the above problems with an evolutionary algorithm and policy gradients method. In this method, agents explore multiple environments with a diverse set of neural networks to train a general model with evolutionary algorithm and policy gradients method. In the experiments, we assume multiple 3D control source tasks. After the model training with our method on the source tasks, we show how effective the model is for the 3D control tasks of the target tasks.

Keywords Deep reinforcement learning · Neuro-evolution · Multi-task learning

---

S. Imai (✉) · Y. Sei · Y. Tahara · R. Orihara · A. Ohsuga
The University of Electro-Communications, Tokyo, Japan
e-mail: imai.shota@ohsuga.lab.uec.ac.jp

Y. Sei
e-mail: seiuny@uec.ac.jp

Y. Tahara
e-mail: tahara@uec.ac.jp

R. Orihara
e-mail: ryohei.orihara@ohsuga.is.uec.ac.jp

A. Ohsuga
e-mail: ohsuga@uec.ac.jp

R. Lee (ed.), *Big Data, Cloud Computing, and Data Science Engineering*,
Studies in Computational Intelligence 844,
https://doi.org/10.1007/978-3-030-24405-7_2

1 Introduction

Deep reinforcement learning that combines reinforcement learning with deep neural networks has been remarkably successful in solving many problems such as robotics [1, 18, 24, 25] and games [6, 19, 27, 33]. Deep reinforcement learning is a method that uses deep neural networks as a function approximator and outputs actions, values and policies.

Deep reinforcement learning needs a lot of samples collected by exploration to training. However, if the amount of exploration space is too many to explore, it takes significant time to collect desired samples and also it is difficult to converge when the exploration is insufficient. In the case of performing a explore using an actual machine in the real world, it is difficult to perform efficient searches due to physical restriction. Also, conducting a search with a policy for which learning has not been completed may cause dangerous behavior for the equipment.

In order to solve this problem, it is necessary to acquire transferable parameters of a neural network by learning on source tasks to make a general model that can learn with small samples on target tasks.

When there are tasks (target tasks) that we want to solve with deep reinforcement learning, we assume that we have other tasks (source tasks) similar to target tasks. If those source tasks are easy to learn by some reason (simple task, simple learning in simulator), it is a likely that parameters common to both tasks can be efficiently acquired by training on the source tasks. In addition, if there are multiple source tasks, by learning parameters that demonstrate high performance for all of these tasks, it is possible to learn good parameters common to tasks in a wide range of source tasks and target tasks.

Existing research has proposed pretraining methods for learning parameters that enable fast learning in multiple tasks by using gradient descent [8]. However, these methods are still limited by several problems, such as the difficulty of learning when the reward is sparse in the pretraining environments [36], deviation of samples [15], and dependence on initial parameters [22].

In this paper, we propose a hybrid multi-task pretraining method by combining an evolutionary algorithm and gradients descent method, that can solve the above problems. In this research, we use Deep Deterministic Policy Gradients (DDPG) [26] as a deep reinforcement learning algorithm to apply our method to 3D continuous control tasks. After the model training with our method in the source tasks, we shows how effective the model is for the 3D control tasks of the target tasks.

This paper is organized as follows. In Sect. 2, we describe the outline of deep reinforcement learning. In Sect. 3, we present related works and the position of our method. In Sect. 4, we detail our proposed algorithm. In Sect. 5, we evaluate our method in experiment and discuss the result of the experiment. Section 6 concludes the paper.

2 Deep Reinforcement Learning

In deep reinforcement learning, deep neural networks are used as a function approximator of a value function or policy. If we use linear function approximator, the convergence of the Q-values is guaranteed, but if the function approximator is nonlinear such as neural networks, convergence is not guaranteed. On the other hand, deep neural networks have high function approximation performance [17]. By utilizing the high approximation performance and feature extraction capability of deep neural network, estimation of effective value function can be expected. In the deep reinforcement learning, the state of the environment is used as an input to deep neural network, and action, state value, the policy, etc are output.

In DQN (Deep-Q-Network) [28] proposed by DeepMind and used on Atari 2600 [2], the state input to the deep neural network is raw frames. Therefore, as a model of deep neural network, a convolutional neural network (CNN) [23] used in image recognition is used. The number of outputs of DQN are the same as the type of actions taken by the agent, and it represents the Q-value of each action. The agent inputs its own observation from environment s_t into DQN and selects the action a_t that has the highest Q-value to play the game.

In learning DQN, we update parameters of neural networks θ for minimizing the following objective function $J(\theta)$:

$$J(\theta) = E[(y_t - Q(s_t, a_t; \theta))^2] \tag{1}$$

$Q(s_t, a_t; \theta)$ is the output by DQN and y_t is the target value to be output. Here, the gradients of parameters θ is as follows.

$$\nabla J(\theta) = 2E[(y_t - Q(s_t, a_t; \theta))\nabla Q(s_t, a_t; \theta)] \tag{2}$$

we update the parameters of DQN according to this gradients.

2.1 *Components of Deep Reinforcement Learning*

The learning of deep reinforcement learning is unstable because deep neural networks with enormous numbers of parameters are used as nonlinear approximators. Therefore, in the DQN method proposed by DeepMind, the following components are used to efficiently perform learning:

1. Target Network
2. Experience Replay

In the loss function used for updating the network, the Q-value after the state following the input state is used as the target value. In order to obtain this target value, if we use same network used to predict the current Q-value to predict the target, training is not stable. Therefore, as a network for outputting the Q-value as the target, the parameters of the Q-network for predicting the Q-value are periodically copied to the target network. As a result, the time lag is established in each network and learning is stabilized. There is another algorithm that improve target network called Double DQN [13].This method is based on the Double Q-learning [12] and generalized to DQN to reduce overestimating action values.

For updating the neural network, we use samples collected by exploring the environment. However, if these samples are input in the order in which they were obtained, the gradient update is performed while ignoring the past experience due to the time series correlation between the samples. To solve this problem, we use a technique called Experience Replay. The samples collected by exploring the environment by the agent are stored in a buffer called a replay buffer. when updating the neural network, sampling is performed randomly from this buffer to make a mini batch, and the neural network is optimized by gradient descent of the parameters using the mini batch as inputs. By using this method, bias due to time series correlation of samples is prevented. In the other version of experience replay that called Prioritized Experience Replay [32], the samples in the buffer are prioritized based on the temporal-difference(TD) error for agent to learn more effectively.

3 Related Works

3.1 *Deep Deterministic Policy Gradients*

DDPG is a deep reinforcement learning algorithm that outputs deterministic policy, represented as the parameters of neural network and optimized these parameters by using policy gradients to maximize expected rewards sum.

In Q-learning [38], when the action space is continuous, it is difficult to find actions with the highest Q-value in a specific state. On the other hand, since DDPG outputs one value deterministically at the output of each action against the input, it is mainly used for the task when the action space is continuous. The DDPG architecture has the actor that outputs action values of each action against the input of observation and has the critic that output the value of action input by the actor by using the output of actor and input of observation. The critic is trained by general supervised learning, and actor uses the critic 's output to learn deterministic policy by using the method called Deterministic Policy Gradients (DPG) [34]. Critic Q with the parameter θ^Q uses the sample (s_i, a_i, r_i, s_{i+1}) from the replay buffer to minimize the following loss function:

$$L = \frac{1}{N}\Sigma_i(y_i - Q(s_i, a_i|\theta^Q))^2 \tag{3}$$

where y_i is derived using the critic's target network Q' with parameters $\theta^{Q'}$, the actor's target network with parameters $\theta^{\pi'}$, discount factor γ.

$$y_i = r_i + \gamma Q'(s_{i+1}, \pi'(s_{i+1}|\theta^{\pi'})|\theta^{Q'}) \quad (4)$$

The actor is trained using the output of critic.

$$\nabla_{\theta^\pi} J \approx \frac{1}{N} \Sigma_i \nabla_a Q(s, a|\theta^Q)|_{s=s_i, a=a_i} \nabla_{\theta^\pi} \pi(s|\theta^\pi)|_{s=s_i} \quad (5)$$

3.2 Meta-learning

MAML (Model-Agnostic Meta-Learning) [8] is a pretraining method applicable to multiple tasks in deep learning. This is a method of obtaining initial parameters of model, which can be used for multiple tasks such as image recognition and reinforcement learning. This method is a kind of method called meta-learning that learns how to learn.

3.3 Multi-task Deep Reinforcement Learning Using Knowledge Distillation

Distal (Distill transfer learning) [37] is a multi-task learning method of deep reinforcement learning. In this method, policies common to individual tasks are measured by KL divergence and extracted into a model by distillation [16]. The extracted policy to a model is transferred to individual tasks and used for each task. Distillation is a method that uses a small neural network (Student Model) to mimic the output knowledge (Dark Knowledge) of another large neural network (Teacher Model). It is used to reduce the number of parameters of the large model. It is shown that simultaneous solving of multiple tasks using this method improves convergence speed of training and improves performance in some tasks.

3.4 Network Exploration Using Evolutionary Algorithm

The method of evolutionary algorithms to explore optimal structure of neural networks is called neuro-evolution [9]. A typical neuro-evolution method is NEAT (Neuro Evolution of Augmenting Topologies) [35], in which this method, we change the structure such as the connections of layers in neural networks to search the optimal network structure by evolutionary computation. There is also a method to play Atari 2600 using NEAT [14]. It is able to surpass the performance of humans in several games. Also, like the proposed method in this paper, PathNet [7] is a method for making a network that can be applied to multiple tasks using evolutionary computation. In this method, embedded agents in the neural network discover which parts of the neural network to re-use for new tasks. In the reinforcement learning tasks of Atari 2600, the pretrained model using this method shows higher performance than the general pretraining method and random initialized model.

3.5 Combining Gradients Descent Methods with Evolutionary Algorithm

As effective learning in deep reinforcement learning, Evolutionary Reinforcement Learning (ERL) [20] and CEM-RL [30] that combines the evolutionary computation and gradients descent method has been proposed. In ERL, the neural network trained by gradients descent is mixed among the population of neural networks trained by the population-based approach of evolutionary computation. Periodically, the parameters of the network trained by gradient descent are copied into an evolving population of neural networks to facilitate training for one task. CEM-RL is a reinforcement learning method that combines the Cross-Entropy Method (CEM) [29] and TD3 [11], the improved algorithm of DDPG. CEM is a kind of evolution strategies algorithm [31]. In this method, a population of actor of new generation is sampled from the mean and the variance of elite individuals of current generation. In CEM-RL, half of actors are directly evaluated and other half of actors are evaluated after updating using policy gradients. After the actors are evaluated, the parameters of next generation are sampled using CEM. In the evolutionary algorithm, since many neural networks are trained simultaneously, it is likely that the most stable neural network is optimized and also this method has characteristics of being robust against the initial parameters of neural networks. In addition to these, since exploration is performed by a plurality of individuals, there is an advantage that it is easy to acquire a reward even if the reward is sparse. ERL is a method that speeds up the task specialization learning using gradients descent learning while solving the problems in reinforcement learning by using an evolutionary algorithm.

3.6 Common Points of Existing Research

The common point of the above multi-task method is that it prevents catastrophic forgetting [10] in training of neural networks. Catastrophic forgetting is the inevitable feature of deep learning that occurs when trying to learn different tasks. As a result of training for one task, information of old tasks is lost. This is an obstacle to making general model that can be used for multiple tasks. In the above research case, it is possible to prevent this by knowledge distillation, meta-learning, and evolutionary algorithm.

3.7 *Position of This Research*

For the tasks in which the reward is sparse because of the enormous action and state space like the 3D-control tasks, there is a problem that the exploration for pretraining itself is difficult [36]. Also, training using a single neural network tends to be influenced by the initial parameters of the neural network [22]. In addition, if we use a single neural network to explore, biasing occurs in the obtained samples, and learning may not be successful due to falling into a local optimum [15]. These problems are independent from the question of "whether or not to get transferable parameters" in the general pretraining method. Therefore, even if there is a methods that can overcome catastrophic forgetting, there is a possibility that pretraining may be prevented by these problems. These problems can be solved by using an evolutionary algorithm like ERL mentioned above. Therefore, by using an evolutionary algorithm like ERL to optimize neural network to multi-tasks, parameters that are considered usable in different tasks can be acquired while overcoming above problems. In this research, by combining optimizing by evolutionary algorithm and gradients descent method on the source tasks to pretraining, we acquire common parameters of the neural network for target tasks to train small exploration in 3D-control tasks.

4 Algorithm

The problem setting of this method is as follows. In our method, population of neural networks is trained on a set of tasks. Among the set of tasks, what is used for pretraining of the model is called source task. Among the set of tasks, new tasks these are not given at the time of pretraining and to be solved after pretraining are called target task. In other words, in this research, the goal of the method is to make a pretrained model that can solve new target tasks using small number of samples to policy gradients after training a model on source tasks.

4.1 *Details of Our Methods*

4.1.1 Exploration by Actors

Figure 1 shows the outline of our method and Algorithm 1 shows the details of our method. First, population of k (k means the number of actors) neural networks (actors) is initialized with random parameters for evolutionary computation. In this research, actors inputs the action value against the input state. Let T be the set of source tasks and each actor π records the reward r_π stored for all tasks in each generation. In addition, the sample (s_i, a_i, r_i, s_{i+1}) sampled in this exploration is

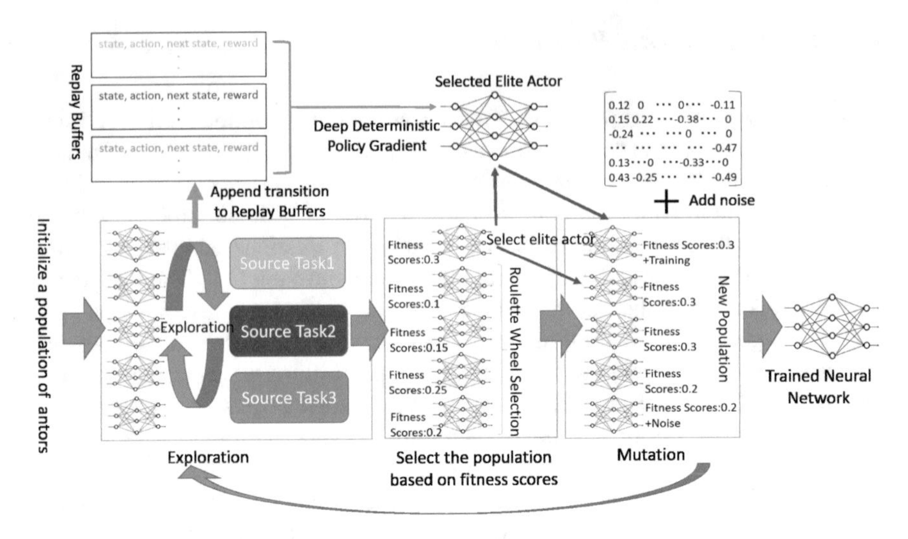

Fig. 1 Multi-task deep reinforcement learning using evolutionary algorithm and policy gradients

---

Algorithm 1 Pseudo-code of Our Method

1: Initialize a population of k actors pop_π with weight θ^π respectively
2: Initialize critic Q with weight θ^Q
3: Initialize replay buffers R
4: Define a random number generator $r() \in [0, 1)$
5: **while** true **do**
6:   **for** actor $\pi \in pop_\pi$ **do**
7:     **for all** source tasks T_i **do**
8:       Explore T_i using θ^π
9:       Append transition to replay buffer R respectively
10:     **end for**
11:   **end for**
12:   Select the elite actor π based on fitness score f_π
13:   Select the replay buffer R based on all fitness scores f_π
14:   Sample a random minibatch of N transitions (s_i, a_i, r_i, s_{i+1}) from R
15:   Update Q by minimizing the loss
16:   $L = \frac{1}{N}\Sigma_i(y_i - Q(s_i, a_i|\theta^Q))^2$
17:   Update copied elite actor π using the sampled policy gradient
18:   $\nabla_{\theta^\pi} J \approx \frac{1}{N}\Sigma_i \nabla_a Q(s, a|\theta^Q)|_{s=s_i,a=a_i} \nabla_{\theta^\pi}\pi(s|\theta^\pi)|_{s=s_i}$
19:   Select the $(k-2)$ actors based on fitness scores f_π and insert selected actors into the next generation's population pop'_π
20:   **for** $(k-2)$ actors $\in pop'_\pi$ **do**
21:     **if** $r() < mut_{prob}$ **then**
22:       Add noise to θ^π
23:     **end if**
24:   **end for**
25:   Insert the elite actor into pop'_π
26:   Insert the copied elite actor pop'_π
27:   $pop_\pi \leftarrow pop'_\pi$
28: **end while**

---

stored in the replay buffers of each tasks. This mechanism makes neural networks to train using unbiased samples by explorations of multiple actors. These samples are used for training actors by the DDPG algorithm.

4.1.2 Elite Selection by Adaptability, Learning

After finishing the exploration with all actors π, the fitness score f_π of each actors π is computed based on the total of recorded rewards, and the actor with the highest fitness score is selected as an elite. The selected elite passes the noise addition after this. To train the elite actor with policy gradients methods, the copy of the selected elite actor is trained by DDPG using samples from a certain task's replay buffer R selected stochastically. Here, the selection probability P_{R_i} of each task's replay buffers R_i is derived using the reward r_i stored by each actors of current generation at each source task T_i.

$$P_{R_i} = \frac{r_i}{\Sigma_j r_j} \tag{6}$$

4.1.3 Selection of Actors Based on Fitness Scores, Adding Noise and Final Selection

After selecting an elite actor, we select actors to be the population of the next generation by roulette selection based on fitness score. After the selection of actors for the next generation, we mutate these networks by adding noise stochastically. Then, next generations actors: An Elite actor, an copied elite actor trained by policy gradients, and selected actors except for elite are copied to the current generation pop_π. The above procedure is repeated until the final generation, and the elite actor in the final generation is an objective neural network.

5 Experiments

In the experiment, we evaluate the performance of the neural networks trained by our pretraining method on the 3D continuous control tasks of OpenAI Gym [3], which is provided by the Pybullet [4].

5.1 *Details of the Experiment*

We evaluate the performance on 6 continuous control tasks (Fig. 2). These tasks are very challenging due to high degree of freedom. In addition, a great amount of exploration is needed to get sufficient reward [5]. Each task has the state such as the

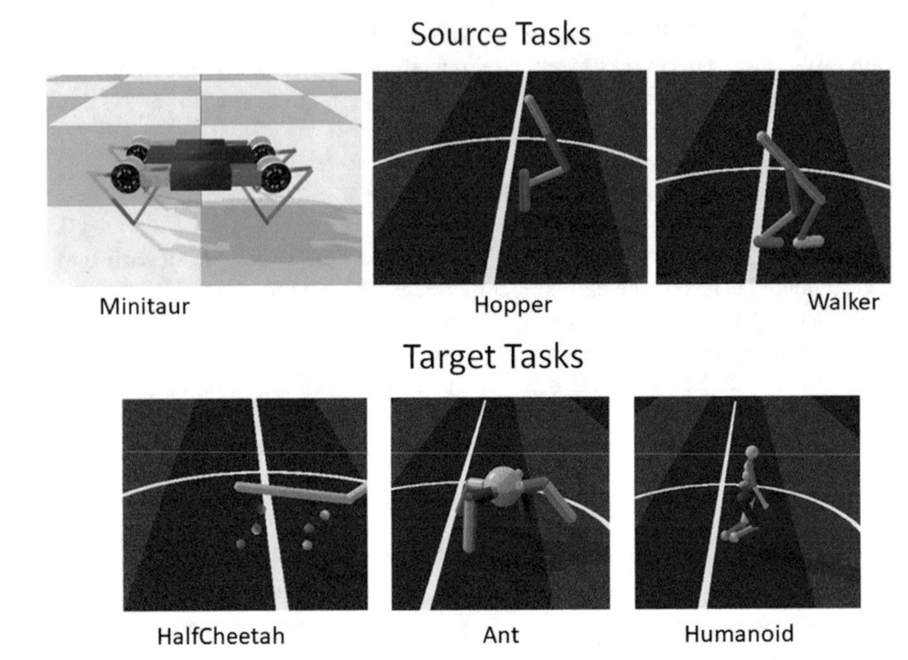

Fig. 2 Source tasks and target tasks used in experiment

value angle of joint, velocity of agents, and output of actions control these values. When agents are controlled properly, we get a reward from the environment. In the source task, actor networks are trained on source tasks through several generations explorations using our method. The selected elite actor in the final generation is set as a pretrained model to be used for all target tasks.

The number of actors in each generation is 10. The maximum number of explorations in one environment is 1000. In the noise-adding phase, we add gaussian noise to actors. The mutation probability of each actor is set to 0.4 in each generation. In order to prevent excessive fluctuation of parameters of the actors, more 0.5 or less −0.5 values among these noises are treated as 0. The number of generations of pretraining in this experiment is 100. Figure 3 shows the network architecture in the experiment. The neural networks used as actors consists of three hidden layers with 128 nodes. The neural networks used as a critic that input observation consists of one hidden layer with 200 nodes, critic that input actor's output and observation neural network consists of one hidden layer with 300 node and the the number of output of critic is one. We use ReLU as the activation function. We use Adam [21] as the optimizer.

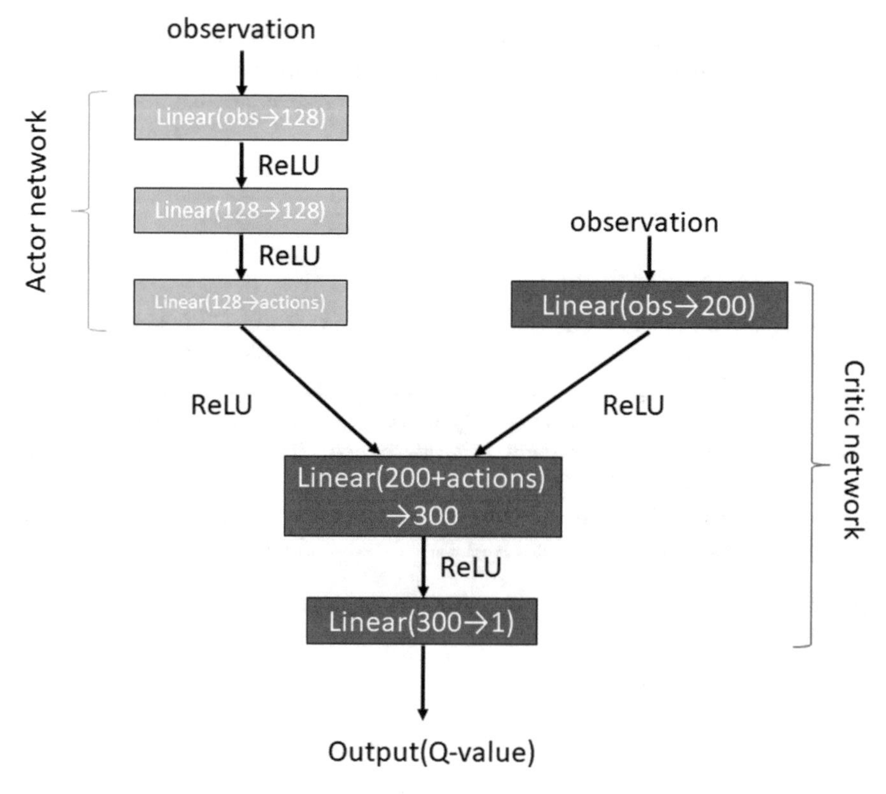

Fig. 3 Network architecture in the experiment

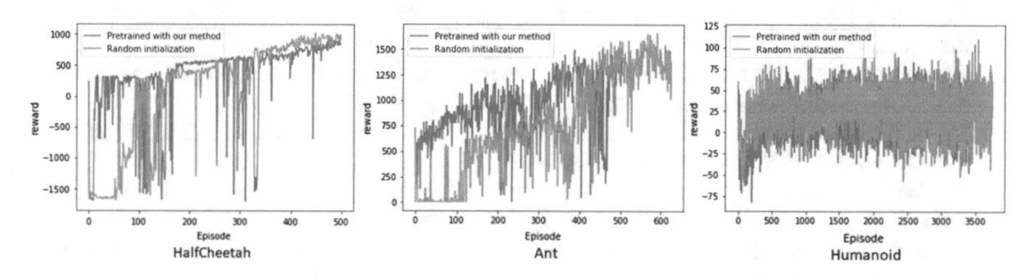

Fig. 4 Training process of the pretrained model and randomly initialized model

5.2 *Results and Discussion*

Figure 4 shows the training process of the pretrained model and randomly initialized model. The horizontal axis indicates the number of episodes and the vertical axis indicates the reward acquired at each episode. In Ant and HalfCheetah tasks, the results show that the amount of reward acquired in the early stage of the episode by the pretrained model exceeds the amount of reward by the randomly initialized model. On the other hand, in the Humanoid task, we can see that the pretrained model cannot acquire efficient reward in the early stage of episode. Since Humanoid task requires observation and action of many parts that are different from other tasks, it may not be possible to learn good parameters in the source task and that the size of the neural network may be insufficient.

6 Conclusion

In this paper, we proposed the method that combines an evolutionary algorithm and policy gradient descent method to deal with the problem that is not solved in the existing pretraining method in deep reinforcement learning. In experiments, we applied the proposed method to the Open AI Gym 3D control task provided by PyBullet, and showed that the neural network trained with our method can adapts to multiple target tasks quickly.

In the future, we would like to confirm the effectiveness of our method by applying it to the problem that the exploration space is larger and reward is sparse.

Acknowledgements This research was supported by JSPS KAKENHI Grant Numbers 16K00419, 16K12411, 17H04705, 18H03229, 18H03340.

References

1. Andrychowicz, M., Baker, B., Chociej, M., Józefowicz, R., McGrew, B., Pachocki, J.W., Petron, A., Plappert, M., Powell, G., Ray, A., Schneider, J., Sidor, S., Tobin, J., Welinder, P., Weng, L., Zaremba, W.: Learning dexterous in-hand manipulation. CoRR arXiv:1808.00177 (2018)
2. Bellemare, M.G., Naddaf, Y., Veness, J., Bowling, M.: The arcade learning environment: an evaluation platform for general agents. CoRR arXiv:1207.4708 (2012)
3. Brockman, G., Cheung, V., Pettersson, L., Schneider, J., Schulman, J., Tang, J., Zaremba, W.: OpenAI gym. CoRR arXiv:1606.01540 (2016)
4. Coumans, E., Bai, Y.: Pybullet, a python module for physics simulation for games, robotics and machine learning (2016–2019). http://pybullet.org
5. Duan, Y., Chen, X., Houthooft, R., Schulman, J., Abbeel, P.: Benchmarking deep reinforcement learning for continuous control. CoRR arXiv:1604.06778 (2016)
6. Espeholt, L., Soyer, H., Munos, R., Simonyan, K., Mnih, V., Ward, T., Doron, Y., Firoiu, V., Harley, T., Dunning, I., Legg, S., Kavukcuoglu, K.: IMPALA: scalable distributed deep-RL with importance weighted actor-learner architectures. CoRR arXiv:1802.01561 (2018)
7. Fernando, C., Banarse, D., Blundell, C., Zwols, Y., Ha, D., Rusu, A.A., Pritzel, A., Wierstra, D.: Pathnet: evolution channels gradient descent in super neural networks. CoRR arXiv:1701.08734 (2017)
8. Finn, C., Abbeel, P., Levine, S.: Model-agnostic meta-learning for fast adaptation of deep networks. In: Precup, D., Teh, Y.W. (eds.) Proceedings of the 34th International Conference on Machine Learning, Proceedings of Machine Learning Research, vol. 70, pp. 1126–1135. PMLR, International Convention Centre, Sydney, Australia (2017). http://proceedings.mlr.press/v70/finn17a.html
9. Floreano, D., Durr, P., Mattiussi, C.: Neuroevolution: from architectures to learning (2008)
10. French, R.M.: Catastrophic forgetting in connectionist networks. Trends Cogn. Sci. **3**, 128–135 (1999)

11. Fujimoto, S., van Hoof, H., Meger, D.: Addressing function approximation error in actor-critic methods. CoRR arXiv:1802.09477 (2018)
12. Hasselt, H.V.: In: Lafferty, J.D., Williams, C.K.I., Shawe-Taylor, J., Zemel, R.S., Culotta, A. (eds.) Advances in Neural Information Processing Systems, vol. 23, pp. 2613–2621. Curran Associates, Inc. (2010). http://papers.nips.cc/paper/3964-double-q-learning.pdf
13. Hasselt, H.V., Guez, A., Silver, D.: Deep reinforcement learning with double Q-learning. In: Proceedings of the Thirtieth AAAI Conference on Artificial Intelligence, AAAI'16, pp. 2094–2100. AAAI Press (2016). http://dl.acm.org/citation.cfm?id=3016100.3016191
14. Hausknecht, M., Lehman, J., Miikkulainen, R., Stone, P.: A neuroevolution approach to general atari game playing. IEEE Trans. Comput. Intell. AI Games (2013). http://nn.cs.utexas.edu/?hausknecht:tciaig14
15. Henderson, P., Islam, R., Bachman, P., Pineau, J., Precup, D., Meger, D.: Deep reinforcement learning that matters. CoRR arXiv:1709.06560 (2017)
16. Hinton, G., Vinyals, O., Dean, J.: Distilling the knowledge in a neural network. In: NIPS Deep Learning and Representation Learning Workshop. arXiv:1503.02531 (2015)
17. Hornik, K., Stinchcombe, M., White, H.: Multilayer feedforward networks are universal approximators. Neural Netw. **2**(5), 359–366 (1989). https://doi.org/10.1016/0893-6080(89)90020-8
18. James, S., Wohlhart, P., Kalakrishnan, M., Kalashnikov, D., Irpan, A., Ibarz, J., Levine, S., Hadsell, R., Bousmalis, K.: Sim-to-real via sim-to-sim: data-efficient robotic grasping via randomized-to-canonical adaptation networks. CoRR arXiv:1812.07252 (2018)
19. Kapturowski, S., Ostrovski, G., Dabney, W., Quan, J., Munos, R.: Recurrent experience replay in distributed reinforcement learning. In: International Conference on Learning Representations (2019). https://openreview.net/forum?id=r1lyTjAqYX
20. Khadka, S., Tumer, K.: Evolution-guided policy gradient in reinforcement learning. In: Bengio, S., Wallach, H., Larochelle,H., Grauman, K., Cesa-Bianchi, N., Garnett, R. (eds.) Advances in Neural Information Processing Systems, vol. 31, pp. 1196–1208. Curran Associates, Inc. (2018). http://papers.nips.cc/paper/7395-evolution-guided-policy-gradient-in-reinforcement-learning.pdf
21. Kingma, D.P., Ba, J.: Adam: a method for stochastic optimization. CoRR arXiv:1412.6980 (2014)
22. Kolen, J.F., Pollack, J.B.: Back propagation is sensitive to initial conditions. In: Proceedings of the 1990 Conference on Advances in Neural Information Processing Systems 3, NIPS-3, pp. 860–867. Morgan Kaufmann Publishers Inc., San Francisco, CA, USA (1990). http://dl.acm.org/citation.cfm?id=118850.119960
23. Krizhevsky, A., Sutskever, I., Hinton, G.E.: Imagenet classification with deep convolutional neural networks. In: Pereira, F., Burges, C.J.C., Bottou, L., Weinberger, K.Q. (eds.) Advances in Neural Information Processing Systems, vol. 25, pp. 1097–1105. Curran Associates, Inc. (2012). http://papers.nips.cc/paper/4824-imagenet-classification-with-deep-convolutional-neural-networks.pdf
24. Levine, S., Finn, C., Darrell, T., Abbeel, P.: End-to-end training of deep visuomotor policies. J. Mach. Learn. Res. **17**(39), 1–40 (2016). http://jmlr.org/papers/v17/15-522.html
25. Levine, S., Pastor, P., Krizhevsky, A., Quillen, D.: Learning hand-eye coordination for robotic grasping with deep learning and large-scale data collection. CoRR arXiv:1603.02199 (2016)
26. Lillicrap, T.P., Hunt, J.J., Pritzel, A., Heess, N., Erez, T., Tassa, Y., Silver, D., Wierstra, D.: Continuous control with deep reinforcement learning. CoRR arXiv:1509.02971 (2015)
27. Mnih, V., Kavukcuoglu, K., Silver, D., Graves, A., Antonoglou, I., Wierstra, D., Riedmiller, M.: Playing atari with deep reinforcement learning. In: NIPS Deep Learning Workshop (2013)
28. Mnih, V., Kavukcuoglu, K., Silver, D., Rusu, A.A., Veness, J., Bellemare, M.G., Graves, A., Riedmiller, M., Fidjeland, A.K., Ostrovski, G., Petersen, S., Beattie, C., Sadik, A., Antonoglou, I., King, H., Kumaran, D., Wierstra, D., Legg, S., Hassabis, D.: Human-level control through deep reinforcement learning. Nature **518**(7540), 529–533 (2015). https://doi.org/10.1038/nature14236

29. Pieter-Tjerk, Kroese, D.P., Mannor, S., Rubinstein, R.Y.: A tutorial on the cross-entropy method. Ann. Oper. Res. **134**(1), 19–67 (2005). https://doi.org/10.1007/s10479-005-5724-z
30. Pourchot, Sigaud: CEM-RL: combining evolutionary and gradient-based methods for policy search. In: International Conference on Learning Representations (2019). https://openreview.net/forum?id=BkeU5j0ctQ
31. Salimans, T., Ho, J., Chen, X., Sutskever, I.: Evolution strategies as a scalable alternative to reinforcement learning. CoRR arXiv:1703.03864 (2017)
32. Schaul, T., Quan, J., Antonoglou, I., Silver, D.: Prioritized experience replay. CoRR arXiv:1511.05952 (2016)
33. Silver, D., Huang, A., Maddison, C.J., Guez, A., Sifre, L., van den Driessche, G., Schrittwieser, J., Antonoglou, I., Panneershelvam, V., Lanctot, M., Dieleman, S., Grewe, D., Nham, J., Kalchbrenner, N., Sutskever, I., Lillicrap, T., Leach, M., Kavukcuoglu, K., Graepel, T., Hassabis, D.: Mastering the game of Go with deep neural networks and tree search. Nature **529**(7587), 484–489 (2016). https://doi.org/10.1038/nature16961
34. Silver, D., Lever, G., Heess, N., Degris, T., Wierstra, D., Riedmiller, M.: Deterministic policy gradient algorithms. In: Xing, E.P., Jebara, T. (eds.) Proceedings of the 31st International Conference on Machine Learning, Proceedings of Machine Learning Research, vol. 32, pp. 387–395. PMLR, Bejing, China (2014). http://proceedings.mlr.press/v32/silver14.html
35. Stanley, K.O., Miikkulainen, R.: Evolving neural networks through augmenting topologies. Evol. Comput. **10**(2), 99–127 (2002). https://doi.org/10.1162/106365602320169811
36. Sutton, R.S., Barto, A.G.: Introduction to Reinforcement Learning, 1st edn. MIT Press, Cambridge, MA, USA (1998)
37. Teh, Y., Bapst, V., Czarnecki, W.M., Quan, J., Kirkpatrick, J., Hadsell, R., Heess, N., Pascanu, R.: Distral: robust multitask reinforcement learning. In: Guyon, I., Luxburg, U.V., Bengio, S., Wallach, H., Fergus, R., Vishwanathan, S., Garnett, R. (eds.) Advances in Neural Information Processing Systems, vol. 30, pp. 4496–4506. Curran Associates, Inc. (2017). http://papers.nips.cc/paper/7036-distral-robust-multitask-reinforcement-learning.pdf
38. Watkins, C.J.C.H., Dayan, P.: Q-learning. Mach. Learn. **8**(3), 279–292 (1992). https://doi.org/10.1007/BF00992698

Learning Neural Circuit by AC Operation and Frequency Signal Output

Masashi Kawaguchi, Naohiro Ishii and Masayoshi Umeno

Abstract In the machine learning field, many application models such as pattern recognition or event prediction have been proposed. Neural Network is a typically basic method of machine learning. In this study, we used analog electronic circuits using alternative current to realize the neural network learning model. These circuits are composed by a rectifier circuit, Voltage-Frequency converter, amplifier, subtract circuit, additional circuit and inverter. The connecting weights describe the frequency converted to direct current from alternating current by a rectifier circuit. This model's architecture is on the analog elements. The learning time and working time are very short because this system is not depending on clock frequency. Moreover, we suggest the realization of the deep learning model regarding the proposed analog hardware neural circuit.

Keywords Analog electronic circuit · Neural network · AC operation · Deep learning

M. Kawaguchi (✉)
Department of Electrical & Electronic Engineering, Suzuka National College of Technology, Shiroko, Suzuka Mie, Japan
e-mail: masashi@elec.suzuka-ct.ac.jp

N. Ishii
Department of Information Science, Aichi Institute of Technology, Yachigusa, Yagusa-cho, Toyota, Japan
e-mail: ishii@aitech.ac.jp

M. Umeno
Department of Electronic Engineering, Chubu University, 1200 Matsumoto-cho, Kasugai, Aichi 487-8501, Japan
e-mail: umeno@solan.chubu.ac.jp

R. Lee (ed.), *Big Data, Cloud Computing, and Data Science Engineering*, Studies in Computational Intelligence 844,
https://doi.org/10.1007/978-3-030-24405-7_3

1 Introduction

In recent years, multi-layered network models, especially the deep learning model, have been researched very actively. The performance is innovatively improved in the field of pattern/speech recognition. The recognition mechanism is elucidated more and more; self-learning IC chips have also been developed. However, these models are working on a general Von Neumann type computer with the software system.

There are only a few studies that have examined the construction of an analog parallel hardware system using a biomedical mechanism. In this research, we propose the neural network, machine learning model with a pure analog network electronic circuit. This model will develop a new signal device with the analog neural electronic circuit. In the field of neural networks, many practical use models, such as pattern recognition or event prediction, have been proposed. And many hardware implementation models, such as vision sensor or parallel network calculator, have been developed.

1.1 Analog Hardware Neural Network

The main merit of an analog hardware neural network is it can operate a continuous time system, not depending on clock frequency. On the other hand, a digital system operates depending on clock behavior on the basement of a Von Neumann type computer. As a result, advanced analog hardware models were proposed [1, 2]. In the pure analog circuit, one of the tasks is the realization of analog memory, keeping the analog value for a while [3]. The DRAM method memorizes in the capacitor as temporary memory, because it can be achieved in the general-purpose CMOS process [4]. However, when the data value is kept for the long term, it needs the mechanism to maintain the memory data. For example, a refresh process is needed. Capacitor reduces the electric charge with the passage of time. It is easy to recover the electric charge of the capacitor using a refresh process in the digital memory. However, in the case of using analog memory, the analog refresh process is quite difficult compared to the digital refresh process of DRAM. Other memorization methods are the floatage gate type device and magnetic substance memories [5, 6].

1.2 Pulsed Neural Network

Pulsed neural network receives time series pulses as learning data and changes the connection weights depending on the number of pulses. This network can keep the connecting weights of the network after the learning process and the outputs of the signal depends on the input value [7]. This network can change the connecting weights by a pulsed learning signal. However, it needs a long time for learning because many pulses are needed. For example, about no less than 1mS is needed in the case where the pulse interval is 10 μS and 100 pulses are received to the network on the learning process.

1.3 The History of Analog Neural Network

Many research results have been reported in the field of neural network and soft computing. These research fields have been widely developed in not only pattern recognition but also control or forecasting systems. The network structure and learning method of a neural network is similar to the biomedical mechanism. The structure of the basic neural network usually consists of three layers. Each layer is composed of the input, connecting weight, summation, thresholds and output unit. In our previous study, we designed and proposed motion detection system or image processing model using a multi-layered neural network and an artificial retina model.

On the other hand, we construct a pattern recognition machine using variable resistance, operational amplifier. We used CdS cells in the input sensor. However, we have to adjust the resistance value by our hands. Moreover, the capacitor reduces the electric charge with time passage. It needs the analog refresh process. The analog refresh process is quite difficult compared to the digital refresh process of DRAM. In the present study, we proposed a neural network using analog multiple circuits and an operational amplifier. The learning time and working time are very short because this system is not dependent on clock frequency in the digital clock processing. At first we designed a neural network circuit by SPICE simulation. Next we measured the behavior of the output value of the basic neural network by capture CAD and SPICE. Capture is one of the computer-aided design (CAD) systems of SPICE simulation. We compared both output results and confirmed some extent of EX-OR behavior [8, 9]. EX-OR behavior is typically a working confirmation method of a three layered neural network model. This EX-OR behavior makes liner separation impossible, it is suitable data for checking neural network ability.

Moreover, the model which used the capacitor as the connecting weights was proposed. However, it is difficult to adjust the connecting weights. In the present study, we proposed a neural network using analog multiple circuits. The connecting weights are shown as a voltage of multiple circuits. It can change the connecting weights easily. The learning process will be quicker. At first we made a neural network by computer program and neural circuit by SPICE simulation. SPICE means the Electric circuit simulator as shown in the next chapter. Next we measured the behavior confirmation of the computer calculation and SPICE simulation. We compared both output results and confirmed some extent of EX-OR behavior [10].

2 Neural Network Using Multiple Circuit

In our previous study, we used multiple circuits to realize the analog neural network. In the SPICE simulation, the circuit is drawn by CAD, called Capture. After setting the input voltage or frequency, SPICE has some analysis function, AC, DC or transient. At first, we made the differential amplifier circuits and Gilbert multiple circuits toward making the analog neural network. We show the different circuits in Fig. 1. Many circuits take an input signal represented as a difference between two voltages. There circuits all use some variant of the differential pair. Figure 1 shows the schematic diagram of the simple transconductance amplifier with differential pairs. The current mirror formed by Q3 and Q4 is used to form the output current, which is equal to I_1–I_2. The different circuits enhanced to a two-quadrant multiplier. Its output current can be either positive or negative, but the bias current I_b can only be a positive current. V_b, which controls the current, can only be positive voltage. So the circuit multiplies the positive part of the current I_b by the tanh of (V_1–V_2). If we plot V_1–V_2 horizontally, and I vertically, this circuit can work in only the first and second quadrants.

We show the Gilbert multiple circuit in Fig. 2. To multiply a signal of either sign by another signal of either sign, we need a four-quadrant multiplier. We can achieve all four quadrants of multiplication by using each of the output currents from the differential pair (I_1 or I_2) as the source for another differential pair. Figure 2 shows the schematic of the Gilbert multiplier. In the range where the tanh x is approximately equal to x, this circuits multiplies V_1–V_2 by V_3–V_4. And we confirmed the range of the voltage operated excellently. One neuron is composed by connecting weights, summation and threshold function. The product of input signal and connecting weights is realized by multiple circuits. Summation and threshold function

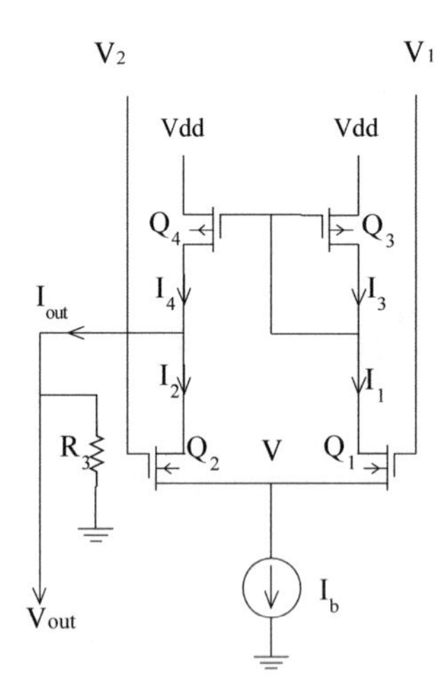

Fig. 1 Difference circuits

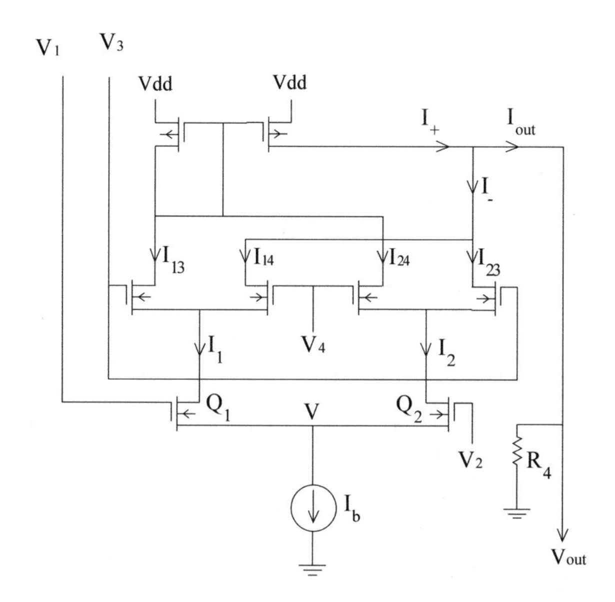

Fig. 2 Gilbert multiple circuits

are realized by additional and difference circuits. In the previous hardware model of neural network, when we use solid resistance elements, it needs to change the resistance elements with the complex wires in each step of learning process. In the case of using variable resistance, we have to adjust the resistance value by our hands.

Figure 3 is the one neuron circuit, using multiple circuits and an additional circuit by opamp. Multiple circuits calculate the product of the two input values, input signals and connecting weights. There are three multiple circuits. Two multiple circuits mean two input signals and connecting weights. The other one multiple circuit means the threshold part of basic neuron. In the threshold part, the input signal is -1. In the multiple circuit, its products input signal -1 and connecting weights. So the output of the multiple circuit is the threshold of this neuron.

3 Perceptron Feedback Network by Analog Circuits

In Fig. 4, we show the architecture of the perceptron neural network. This is a basic learning network using a teaching signal. ‘y’ means the output signal of the neural network. ‘t’ means the teaching signal. Error value ‘t-y’ is calculated by the subtract circuit. After calculating the error value, the product error value and input signal are calculated and make a feedback signal. The input of the subtract circuit on the feedback line are the feedback signal and the original connecting weight. This subtract circuit calculates the new connecting weight. After the product of the new connecting weigh is calculated, the next time learning process is started.

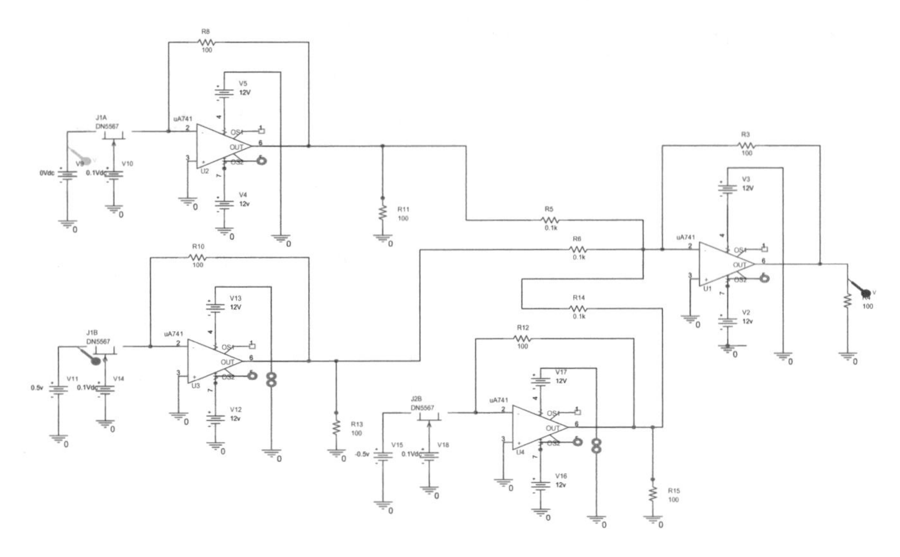

Fig. 3 Neural circuit by capture CAD

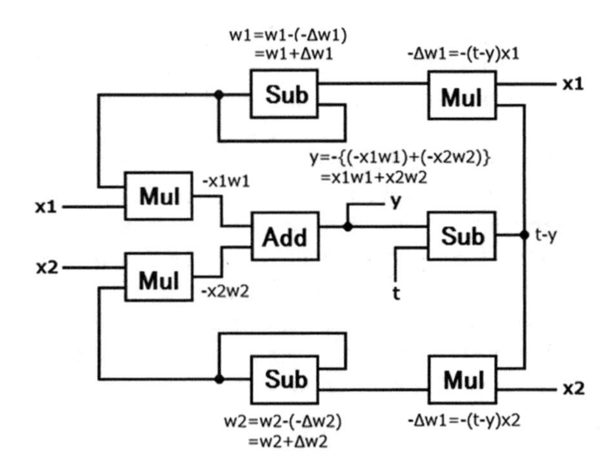

Fig. 4 The architecture of perceptron

Figure 5 shows the perceptron circuits, two-input and one-output. There are multiple circuits and additional circuits in the feed forward line. Error value between original output and teaching signal is calculated by subtract circuit. There are multiple circuits and additional circuit in the feedback lines. In the experimental result of this perceptron, the learning time is about 900 μS shown in Fig. 6 [11]. Figure 7 shows the Architecture of Three-Layers Neural Circuits. In Fig. 8, we show the Learning Neural Circuit on Capture CAD by SPICE.

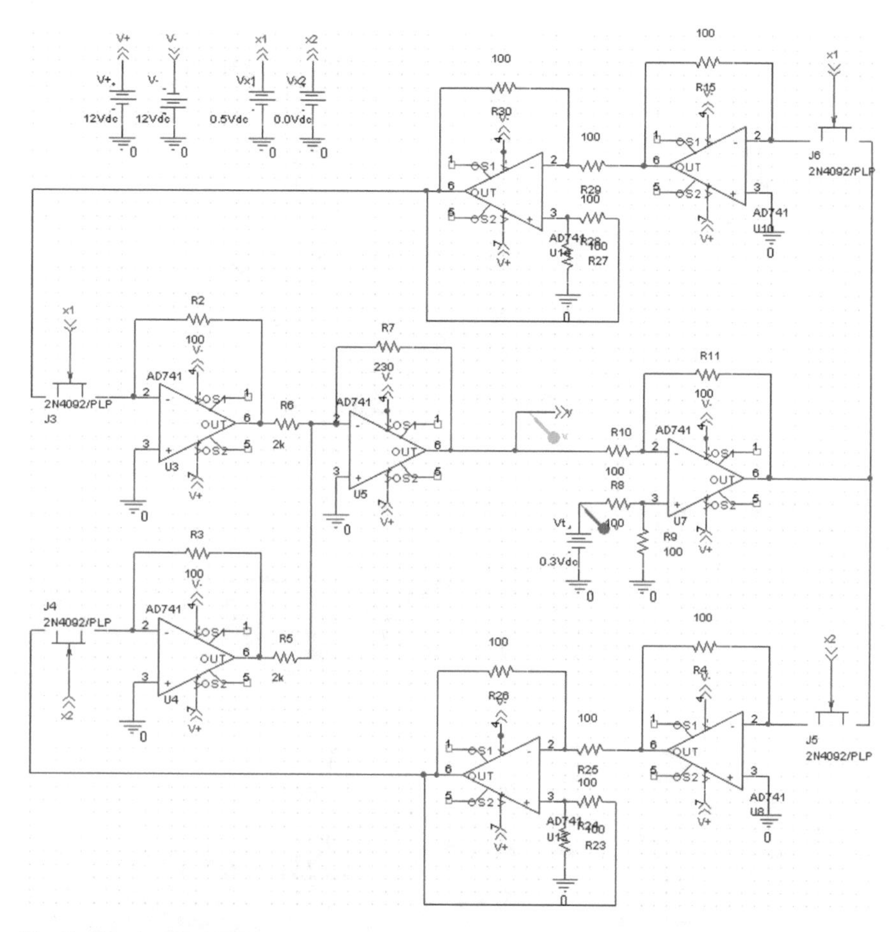

Fig. 5 The circuit of perceptron

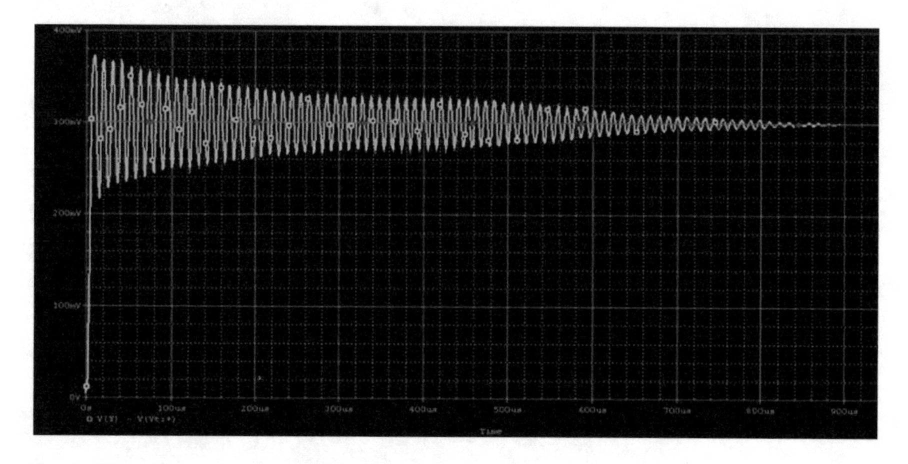

Fig. 6 The convergence output of perceptron

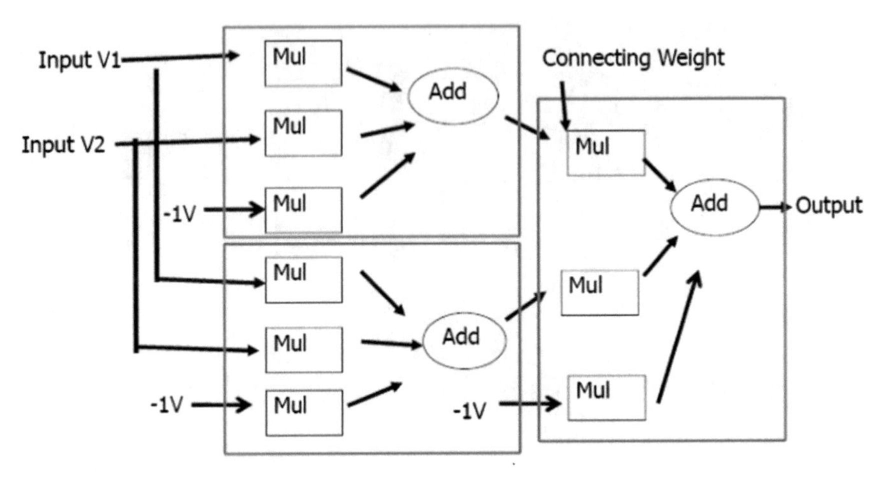

Fig. 7 The diagram of neural circuits with threshold

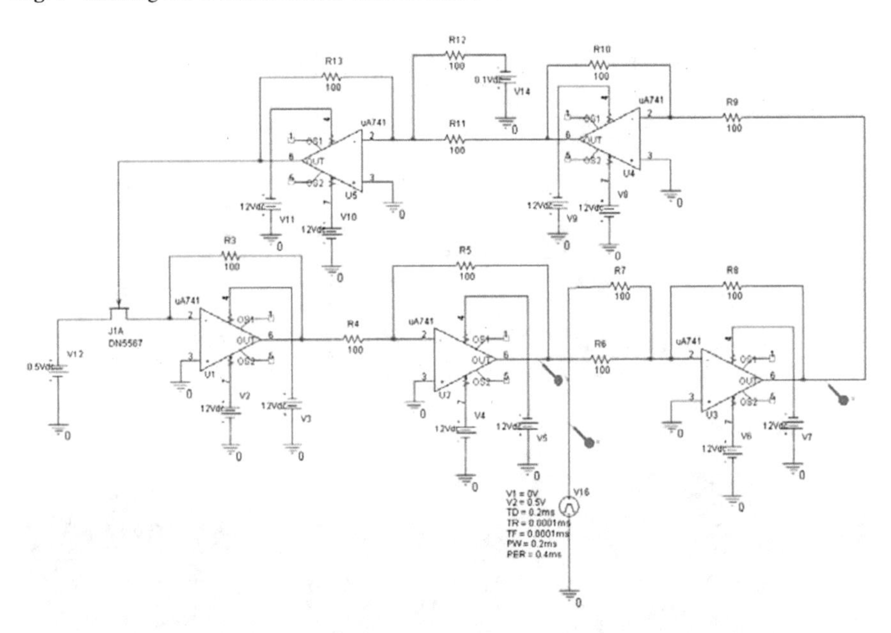

Fig. 8 The learning feedback neural circuit

4 Neural Circuit on Alternative Current Behavior

We proposed an analog neural network using multiple circuit in the previous research. However, in the case of constructing a network, one of the disadvantages is that the input and output range is limited. Furthermore, the circuit operation becomes unstable because of the characteristics of the multiple circuit using semiconductor. It is called 'Circuit Limitations'. One of the cause is transistor mismatch. Not all transistors created equal. Another cause is the output-voltage limitation. We tried to use the alternative current as a transmission signal in the analog neural network in Fig. 9. The original input signal is direct current. We used the voltage frequency converter unit when generate the connecting weight. The input signal and connecting weight generate the Alternative current by the Amplifier circuit. Two Alternative currents are added by an additional circuit. The output of the additional circuit is a modulated wave. This modulated wave is the first output signal of this neural network.

Figure 10 is the output of the RMS value of AC voltage by the neural circuit. In this network, two Alternative currents are added by an additional circuit. The output of the additional circuit is a modulated wave. Figure 10 shows the RMS value of the modulated wave. It operates satisfactorily because the output voltage increases monotonically in the general-purpose frequency range. Figure 11 is the output of the neural circuit. It is shown by two dimensional graph. We recognized the RMS value of the output voltage is the appropriate value in the two-dimensional area.

When we construct learning the AC neural circuit, we have to convert the feedback modulated current signal to a connecting weight with frequency. The correction error signal is calculated by the products difference signal and input signal. Difference signal is the difference between the output value and teaching signal. Figure 12 shows the convergence result of learning experiment. It means the learning process is succeeding with very short time. Figure 13 shows the Basic AC operation learning neural network model. This circuit is composed by a Rectifier circuit, Voltage-Frequency converter, Amplifier, subtract circuit, Additional Circuit and Inverter. The input sig-

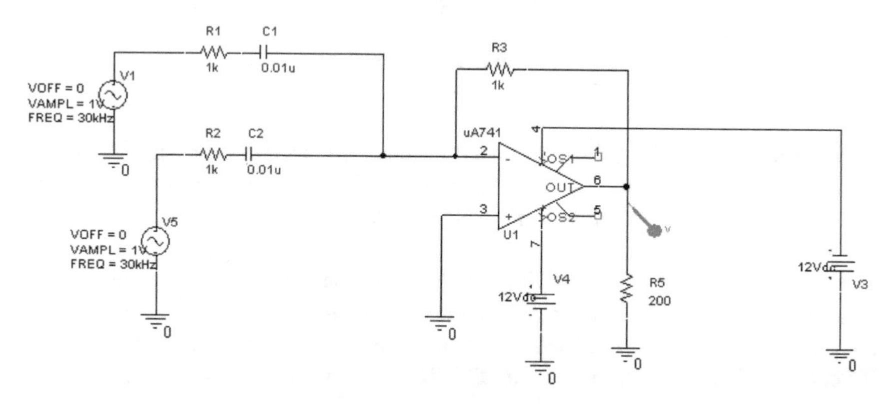

Fig. 9 AC operation neural circuit

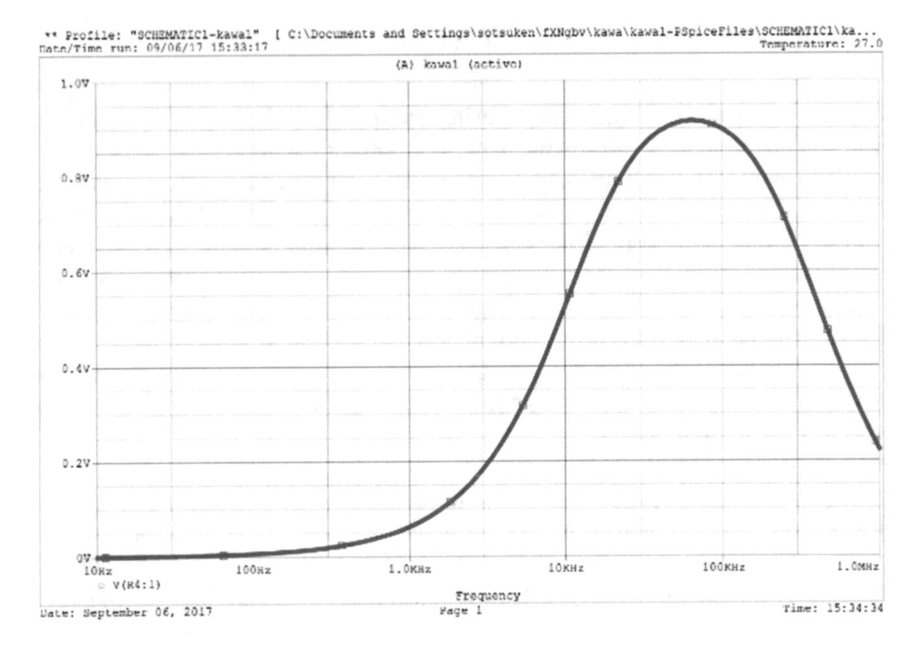

Fig. 10 The output Rms value of neural circuit

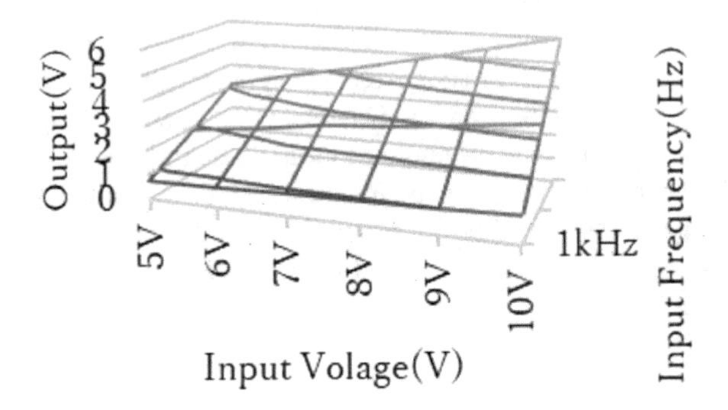

Fig. 11 The output behavior of AC operation neural circuit

nal is direct current. The initial value of the connecting weight is also direct current. This direct current is converted to frequency by a Voltage-Frequency converter circuit. The input signal and connecting weight generate the Alternative current by the Amplifier circuit.

Figure 14 shows the relationship between the number of learning time and output frequency. It shows the frequency f1 convergences to 4 kHz and the frequency f2 convergences to 1 kHz. The learning count time is very small. The learning speed is very fast in this AC operation circuit. Figure 15 shows the whole circuit of AC operation learning neural network.

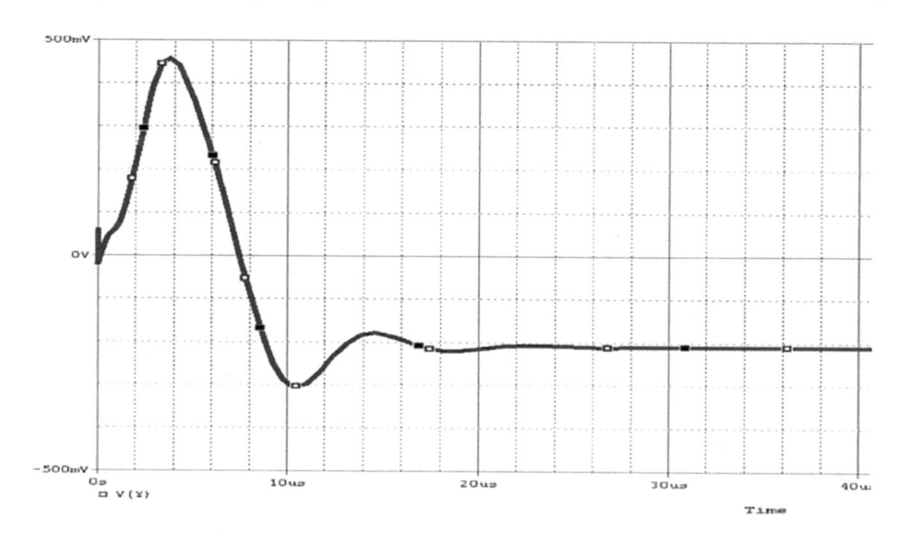

Fig. 12 The convergence result of learning experiment

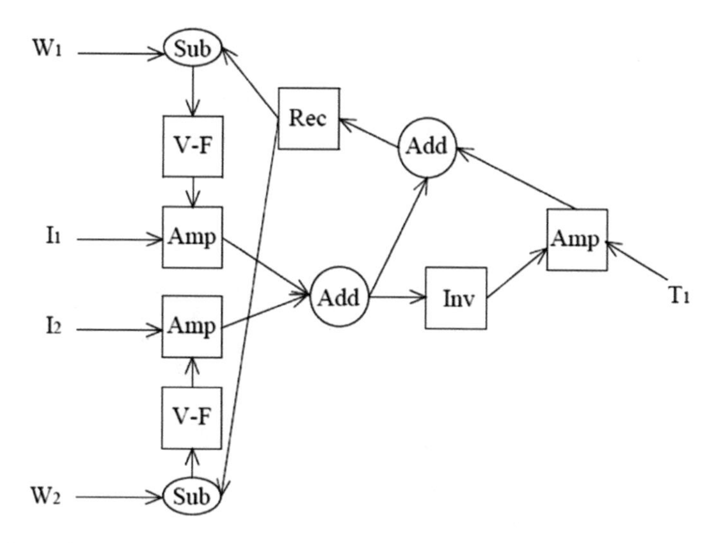

Fig. 13 Basic AC operation learning neural network model

Two alternative currents are added by an additional circuit. The output of the additional circuit is a modulated wave. This modulated wave is phase inverted by an Inverse circuit. The phase-inverted wave is amplified. The amplification is the value of the teaching signal. This amplified signal and modulated wave are added by an adder circuit. The output of this adder circuit is the error value, which is the difference between the output and teacher signal. Thus, we do not have to use the subtract circuit to calculate the error value (Figs. 17 and 18).

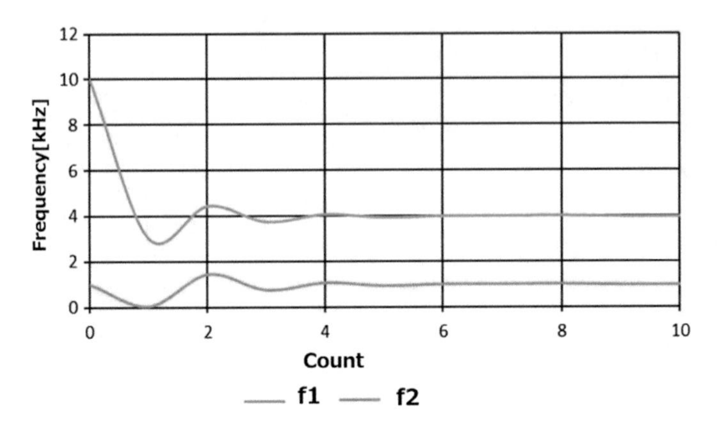

Fig. 14 The number of learning time and output frequency

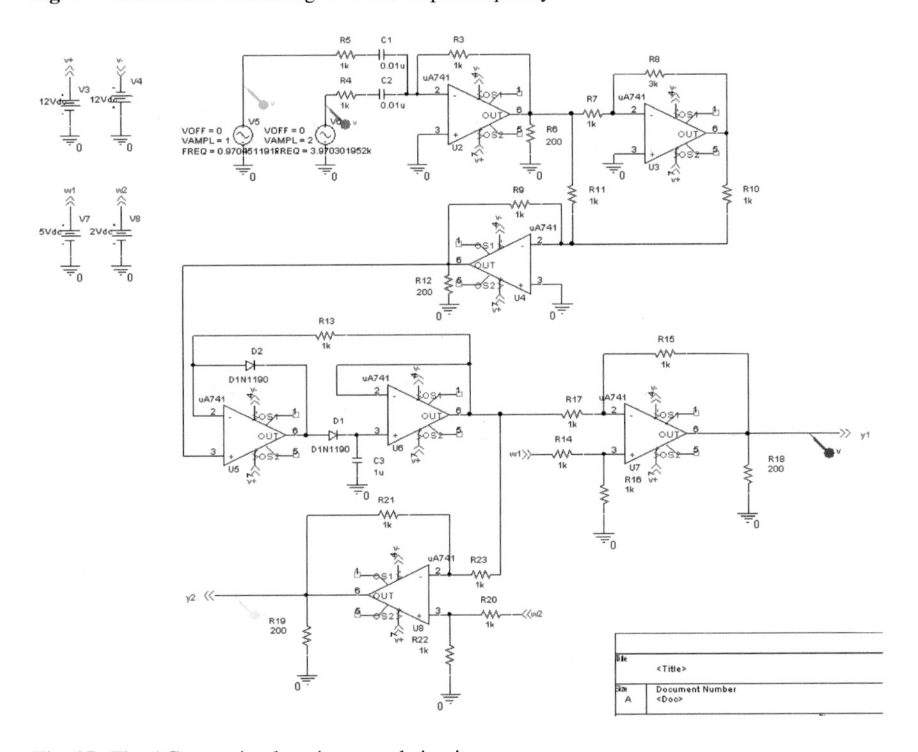

Fig. 15 The AC operation learning neural circuit

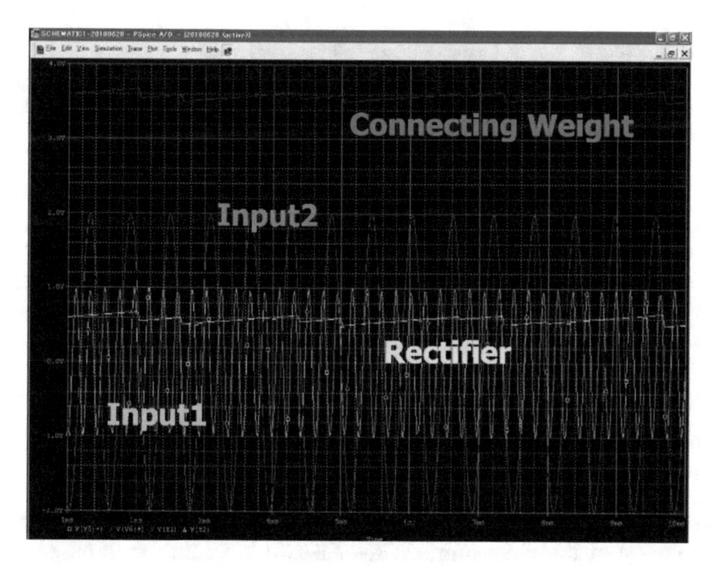

Fig. 16 The simulation results of AC feed-back neural model

The output of the Adder circuit is converted to direct current from alternating current by the rectifier circuit. This direct current is a correction signal of connecting weights. New connecting weight is calculated by a subtract circuit. This circuit calculates the original connecting weight and the correction signal of the connecting weight. The output of the subtract circuit is converted to a frequency signal by a voltage-frequency convert circuit. It means that in the AC feedback Circuit for the BP learning process, after getting DC current by a rectifier circuit, we have to convert from DC Voltage to AC current with frequency. Finally, alternating current occurs by the amplifier circuit. The amplification is the value of the input signal. Figure 16 shows the simulation results of AC feed-back neural model, two-input signal, connecting weights and after rectified wave.

5 Deep Learning Model

Recently, a deep learning model has been proposed and developed in many applications such as image recognition and artificial intelligence. Deep learning is a kind of algorithms in the machine learning model. This model is developed in the recent research. The recognition ability is improved more and more. Not only pattern recognition, but also in image or speech recognition fields, the deep learning model is used in many fields in practical use. And this system is expected in the field of robotics, conversation system and artificial intelligence.

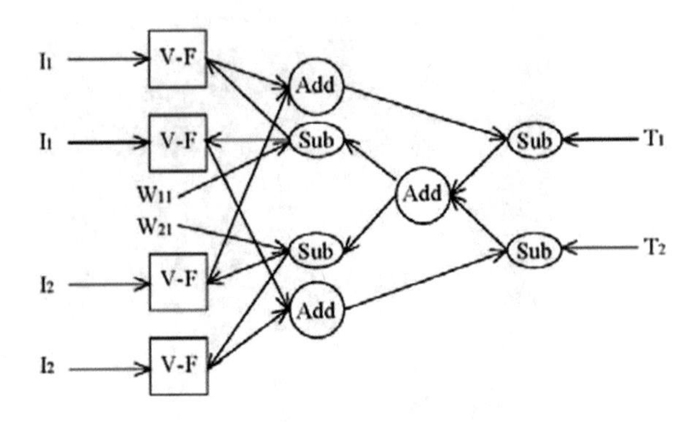

Fig. 17 The structure of 2-patterns analog AC operation neural network with V-F conversion circuit

5.1 The Stacked Auto Encoder

The stacked auto-encoder is one of the processes in deep learning. This is the pre-learning method of the large number layer network. In the basic neural network, in the almost case, there are three layers. On the other hand, there are nine layers in the deep learning model generally. After the learning process is completed by machine learning method, remove the decoding part, output layer and the connection of intermediate layer and output layer. The keeping of the coded portion means from the input layer to the intermediate layer including the connection of the input layer and intermediate layer. The intermediate layer contains the compressed data of the input data. Moreover, we obtain more compressed internal representation, as the compressed representation input signal to apply the auto-encoder learning. After removing the decoding part of stacked auto-encoder, next network is connected. This network is also learned by another three-layered network and remove the decoding part. A Stacked auto-encoder has been applied to the Restricted Boltzmann Machine (RBM) as well as the Deep learning network (DNN). Moreover, stacked auto-encoder is used the many types of learning algorithm. Recently, the learning experiment featuring a large amount of extraction from an image has become well-known. Stacked auto-encoder can self-learning of abstract expression data. This network has nine layers with three superimposed sub-networks, such as a convolution network [12]. In the previous research, we described the simple neural network learning model by analog electronic circuits. We tried to expand the network to realize the deep learning model. Next, we constructed 2 input, 1 output and 2 patterns neural model as in Fig. 14. In this circuit, each pattern needs each circuit. For example, in the case there are 5 kinds of learning patterns, we have to construct 5 input unit circuits. However, learning time is very short. "V-F" means the V-F converter circuit. The output of the subtract circuit is converted to a frequency signal by a voltage-frequency convert circuit in Fig. 17.

The output of the subtract circuit is converted to a frequency signal by a voltage-frequency convert circuit. It means that in the AC feedback Circuit for BP learning process, after getting DC current by rectifier circuit, we have to Convert from DC Voltage to AC current with Frequency. I_1 and I_2 are input units. Two I_1 mean two

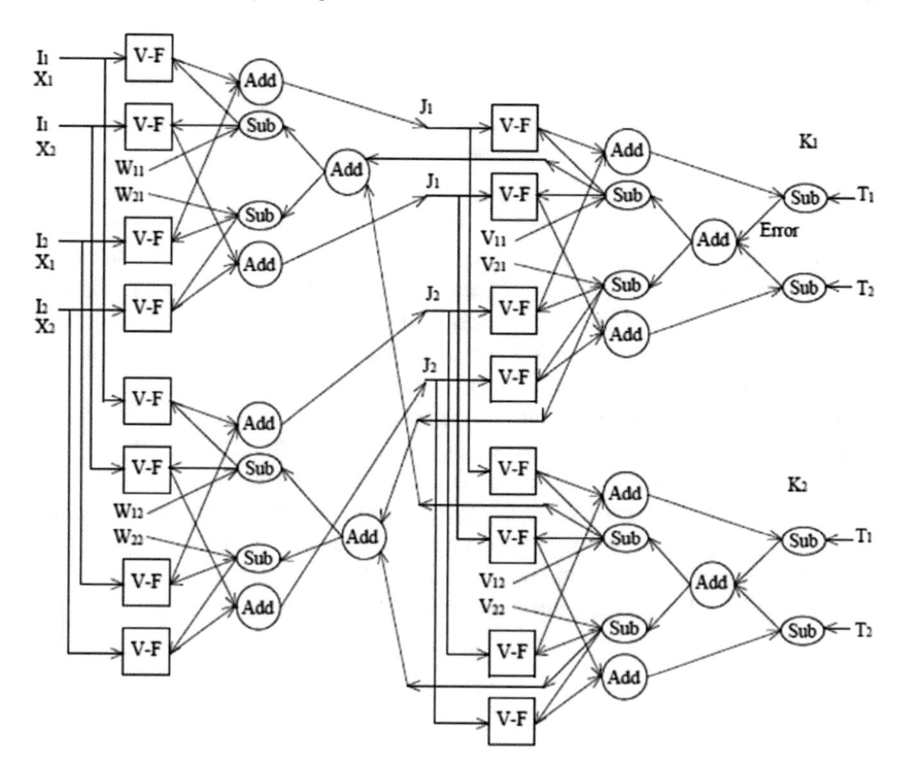

Fig. 18 The structure of enhanced analog neural network to three layers of Fig. 15

inputs. T_1 and T_2 means two teaching signals. W_{11} and W_{12} are connecting weights. Figure 18 means the expand network of Fig. 17. Although this model needs many neural connections, the learning speed is very high because of the plural data patterns learning occurs at the same time and working analog real time system not depending on clock frequency. And after learning, each new connecting weight between the input layer and middle layer is picked up, it is the parted potion including the connecting weights between the input layer and middle layer as well as the layers of input and middle. It means the stacked auto encoder process and suggests the possibility of design of many layers of the deep learning model [13]. To fix the connecting weights after learning process, we proposed the two-stage learning process. In the learning stage, connecting weighs are able to change depending on the teaching signal. After learning process finished, we used the sample hold circuit to fix the connecting weights. In this situation, this circuits receive the input signal and outputs the output signal in the environment that all the connecting weights are fixed.

6 Conclusion

At first, we designed an analog neural circuit using multiple circuits. We confirmed the operation of this network by SPICE simulation. Next, we constructed a basic analog neural network by an alternative current operation circuit. The input signal and connecting weight generate the alternative current by the amplifier circuit. Two alternative currents are added by an additional circuit [14, 15]. The frequency signal is generated by a Voltage-Frequency converter circuit. The input signal of the V-F converter is rectified direct current. The input of the rectified circuit is the error correction signal by alternative current. The connecting weight can be changed by an error-correction signal and the input frequency is depending on the output Voltage-Frequency converter circuit in the feedback learning process. This model has extremely high flexibility characteristics. It is the AC feedback circuit for the BP learning process, after getting DC current by rectifier circuit, we have to convert from DC Voltage to AC current with frequency. Moreover, a deep learning model has been proposed recently and developed in many applications such as image recognition and artificial intelligence. In the future, this hardware learning system is expected in the field of robotics, conversation systems and the artificial intelligence.

References

1. Mead, C.: Analog VLSI and Neural Systems, Addison Wesley Publishing Company, Inc. (1989)
2. Chong, C.P., Salama, C.A.T., Smith, K.C.: Image-motion detection using analog VLSI. IEEE J. Solid-State Circuits **27**(1), 93–96 (1992)
3. Lu, Z., Shi, B.E.: Subpixel resolution binocular visual tracking using analog VLSI vision sensors. IEEE Trans Circ Syst II Anal Digital Signal Process **47**(12), 1468–1475 (2000)
4. Saito, T., Inamura, H.: Analysis of a simple A/D converter with a trapping window. IEEE Int. Symp. Circ. Syst. 1293–1305 (2003)
5. Luthon, F., Dragomirescu, D.: A cellular analog network for MRF-based video motion detection. IEEE Trans Circ Syst II Fundamental Theor Appl **46**(2), 281–293 (1999)
6. Yamada, H., Miyashita, T., Ohtani, M., Yonezu, H.: An analog MOS circuit inspired by an inner retina for producing signals of moving edges. Technical Report of IEICE, **NC99–112**, 149–155 (2000)
7. Okuda, T., Doki, S., Ishida, M.: Realization of back propagation learning for pulsed neural networks based on delta-sigma modulation and its hardware implementation. ICICE Trans. J88-D-II-4, 778–788 (2005)
8. Kawaguchi, M., Jimbo, T., Umeno, M.: Motion detecting artificial retina model by two-dimensional multi-layered analog electronic circuits. IEICE Trans. E86-A-2, 387–395 (2003)
9. Kawaguchi, M., Jimbo, T., Umeno, M.: Analog VLSI layout design of advanced image processing for artificial vision model. In: IEEE International Symposium on Industrial Electronics, ISIE2005 Proceeding, vol. 3, pp. 1239–1244 (2005)
10. Kawaguchi, M., Jimbo, T., Umeno, M.: Analog VLSI layout design and the circuit board manufacturing of advanced image processing for artificial vision model. KES2008, Part II, LNAI, **5178**, 895–902 (2008)
11. Kawaguchi, M., Jimbo T., Ishii, N.: Analog learning neural network using multiple and sample hold circuits. IIAI/ACIS International Symposiums on Innovative E-Service and Information Systems, IEIS 2012, 243–246 (2012)

12. Yoshua, B., Aaron, C., Courville, P.: Vincent: representation learning: a review and new perspectives. IEEE Trans. Pattern Anal. Mach. Intell. **35**(8), 1798–1828 (2013)
13. Kawaguchi, M., Ishii, N., Umeno, M.: Analog neural circuit with switched capacitor and design of deep learning model. In: 3rd International Conference on Applied Computing and Information Technology and 2nd International Conference on Computational Science and Intelligence, ACIT-CSI, pp. 322–327 (2015)
14. Kawaguchi, M., Ishii, N., Umeno, M.: Analog learning neural circuit with switched capacitor and the design of deep learning model. Computat. Sci. Intell. Appl. Informat. Stud. Computat. Intell. **726**, 93–107 (2017)
15. Kawaguchi, M., Ishii, N., Umeno, M.: Analog neural circuit by AC operation and the design of deep learning model. In: 3rd International Conference on Artificial Intelligence and Industrial Engineering on DEStech Transactions on Computer Science and Engineering, pp. 228–233
16. Kawaguchi, M., Jimbo, T., Umeno, M.: Dynamic Learning of Neural Network by Analog Electronic Circuits. Intelligent System Symposium, FAN2010, S3-4-3 (2010)

IoTDoc: A Docker-Container Based Architecture of IoT-Enabled Cloud System

Shahid Noor, Bridget Koehler, Abby Steenson, Jesus Caballero, David Ellenberger and Lucas Heilman

Abstract In recent years, cloud computing has gained considerable notoriety because it provides access to shared system resources, allowing for high computing power at low management effort. With the widespread availability of mobile and Internet-of-Things (IoT) devices, we can now form cloud instantly without considering a dedicated infrastructure. However, the resource-constrained IoT devices are quite infeasible for installing virtual machines. Therefore, the mobile cloud can be used only for the tasks that require distributed sensing or computation. In order to solve this problem, we introduce IoTDoc, an architecture of mobile cloud composed of lightweight containers running on distributed IoT devices. To explore the benefits of running containers on low-cost IoT-based cloud system, we use Docker to create and orchestrate containers and run on a cloud formed by cluster of IoT devices. We provide a detail operational model of IoTDoc that illustrates cloud formation, resource allocation, container distribution, and migration. We test our model using the benchmark program Sysbench and compare the performance of IoTDoc with Amazon EC2. Our experimental result shows that IoTDoc is a viable option for cloud computing and is a more affordable, cost-effective alternative to large platform cloud computing services, specifically as a learning platform than Amazon EC2.

S. Noor (✉)
Northern Kentucky University, Highland Heights, USA
e-mail: noors2@nku.edu

B. Koehler · A. Steenson · J. Caballero · D. Ellenberger · L. Heilman
St. Olaf College, Northfield, USA
e-mail: koehle2@stolaf.edu

A. Steenson
e-mail: steens1@stolaf.edu

J. Caballero
e-mail: caball1@stolaf.edu

D. Ellenberger
e-mail: ellenb1@stolaf.edu

L. Heilman
e-mail: heilma1@stolaf.edu

R. Lee (ed.), *Big Data, Cloud Computing, and Data Science Engineering*,
Studies in Computational Intelligence 844,
https://doi.org/10.1007/978-3-030-24405-7_4

Keywords Docker · Container · Mobile cloud · Ad hoc cloud · IoT

1 Introduction

Cloud computing allows access to shared system resources, resulting in great computing power with relatively small management effort. Virtual machines can be leveraged to make use of the resources of the physical machine without directly interacting with it. Though the traditional cloud system is immensely popular [1] it has a high initial setup and maintenance cost [2]. Moreover, the traditional cloud fails to perform any task that requires distributed sensing or computation in a network-disconnected area [3]. Therefore, several researchers proposed mobile cloud considering loosely connected mobile and IoT devices. A mobile cloud can be formed anywhere instantly and have more flexibility for service selection and price negotiation. However, installing virtual machines in a resource-constrained mobile or IoT devices is really challenging. Therefore, the existing mobile cloud architectures primarily consider running the user specified tasks directly to the distributed devices in the mobile cloud. Since a mobile device owner can accept multiple tasks simultaneously, it is hard to monitor and audit the tasks individually. Moreover, ensuring the safety of devices in a mobile cloud while running multiple tasks is very difficult without the absence of any virtualized platform. Therefore, we need to provide a lighter platform than virtual machines that will be able to create a logical separation of tasks and ensure the safety and security of the devices.

In recent time, several types of research are done for creating a more lightweight OS-level virtualization. Tihfon et al. proposed a multi-task PaaS cloud infrastructure based on docker and developed a container framework on Amazon EC2 for application deployment [4]. Naik proposed a virtual system of systems using docker swarm in Multiple Clouds [5]. Bellavista et al. [6], Morabito [7], and Celesti et al. [8] showed the feasibility of using containers over Raspberry Pis. However, all of the above works did not discuss the formation of containers on an IoT-based cloud system. Though some of the researchers presented high-level architectures for deploying containers on IoT devices [9, 10], those architectures do not have any proper operational model for container creation, cluster formation, container orchestration, and migration. Recently, Google's developed open source project Kubernetees got immense popularity that automates Linux container operations [11]. Though Kubernetes users can easily form clusters of hosts running on Linux containers and manage easily and efficiently those clusters, the implementation of Kubernet for varying size of containers on resource constraint IoT Cloud is quite challenging [12, 13].

Therefore, we present, IoTDock, an architecture of IoT-based mobile cloud that uses Docker engine to create and orchestrate containers in the IoT devices and provides granular control to users using the cloud service. Unlike virtual machines, a container does not store any system programs (e.g., operating systems) and, therefore, uses less memory than the virtual machine and requires less storage. Docker has built-in clustering capabilities that also ensures the security of the containers.

We configure Docker Swarm, a cluster of Docker engines, which allows application deployment services without additional orchestration software. In our operational model, we propose an efficient cluster head selection and resource allocation strategy We also present container image creation and installation procedure. Moreover, we illustrate different scenarios for container migration along with countermeasures. Finally, we design a prototype of IoTDoc and provide a proof of concept by running various diagnostic tests to measure power consumption, financial cost, install time, runtime, and communication time between devices. We create a cluster using Amazon Web Services (AWS) and compare this AWS-based model with our IoTDoc using some sample benchmark problems.

Contributions

- We present a resource allocation strategy. As far as we know this is the first attempt to provide such strategy for a distributed system composed of multiple IoT devices that can receive containers from multiple users.
- We propose a model for efficient container creation and container migration for IoT-enabled distributed system.

The rest of the sections are organized as follows. In Sects. 2 and 3 we discuss the background and motivation associated with our proposed work. We present our conceptual architecture in Sect. 5 followed by a detailed operational model in Sect. 6. We show our experimental result in Sect. 7. Finally, we discuss and conclude our work in Sects. 8 and 9 respectively.

2 Background

The term mobile cloud has been defined as cloud-based data, applications, and services designed specifically to be used on mobile and/or portable devices [2]. Mobile cloud provides distributed computation, storage, and sensing as a services considering the crowdsourced mobile or IoT devices distributed in different parts of the world [14].

One method of application delivery that has garnered considerable attention in recent years is Docker. Docker is a container platform provider that automates the deployment of applications to run on the cloud through container images. A container image is a package of standalone software that consists of everything required for execution including code, system libraries, system tools, runtime, and settings [15].

Docker has grown considerably since its inception in 2013. Currently, over 3.5 million applications have been containerized using Docker. Additionally, these containerized applications have been downloaded over 37 billion times [16]. This rise in the popularity of containerization rivals that of virtual machines (VMs). While VMs and containers have comparable allocation and resource isolation benefits, containers are more lightweight and efficient [17]. In addition to the application and required libraries, VMs also include a full guest operating system, as illustrated in Fig. 1

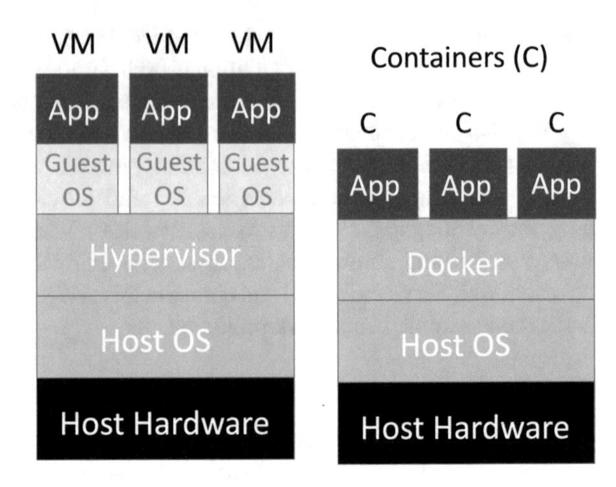

Fig. 1 VM versus container-based model

and require more resources to reach full virtualization than containers do. As a consequence, containers start more quickly, deploy more easily, and allow for higher density than VMs.

3 Motivation

The motivation behind this research can be traced back to the spread of the IoT and related connectable devices. The global IoT market is predicted to expand from $157B in 2016 to $457B by 2020 [18], creating a vast network of connected devices that directly adds to possible computational power. The bulk of this network is not just mini-SoCs like Raspberry Pis, but instead a diverse ecosystem full of devices from coffee makers to cars. In theory, this diversity will continually branch into more and more types of chips, sensors, and devices with varying functions and, in a sense more applicable to our needs, computational power that can be leveraged. If microchips continue to follow Moore's Law [19], future creation of low-cost distributed computing platforms may be just as easy as going out and buying a laptop. However, to leverage all of these mobile devices, a large number need to be pieced together in an efficient manner. To approach such a cluster, we looked at what other researchers have done, settling on using Docker, a distributed hosting framework as our means to manage devices. Docker has been used in previous research with Pi clusters to fair results [6]. Our goal would be to expand on just how feasible this may be today.

Migration is another primary challenge in using a virtual machine for hardware level separation. Researchers presented several approaches to reduce the time associated with service handoff inside an edge network [20]. Most of those approaches determine some important portion and migrate those targeted portion instead of migrating the whole VM. The problem with those proposed approaches are the selected portion of the VM is significantly large in size because of the size of the actual VM. Therefore, using VM for low power IoT enabled edge computing system is inefficient especially for a task migration.

4 Related Work

In recent years the use of both hardware and OS-level virtualization has increased rapidly. Researchers illustrated the advantages of using virtualization with numerous experimental evaluation. Felter et al. [21] compared the overhead of using virtualized technologies, such as KVM and Docker-Container and non-virtualized technologies and showed that Docker containers equal or exceed the performance of KVM systems in almost every scenario. Agarwal et al. [22] showed that the memory footprint and startup times of a container is slightly smaller than a virtual machine. Using experimental evaluation they compared a container-based system with a VM-based system and showed the former has approximately 6 times lower startup time and 11 times lower memory footprint than a VM-based model. Sharma et al. [23] showed that the performance of both VM and Container based system degrades for co-located applications and the result varies depending on the type of workloads. They also showed the effect of different type of settings inside a VM or container on the management and development of applications. Tihfon et al. presented a container-based model for multi-task cloud infrastructure [4] that are able to build an application in any language. They also evaluated their model and showed the overall response and data processing time of a container-based model for multiple service models. Naik presented a virtual system of systems (SoS) based on Docker Swarm for the distributed software development on multiple clouds [5]. All of the above-mentioned works considered designing containers in a traditional stationary host-based infrastructure. In contrast, we present a container-based system for mobile infrastructure where the containers will be installed on distributed mobile and IoT devices. Along with reducing the cost of setup and maintenance, our model can also provide more flexibility for task selection and task migration.

So far a very few researchers addressed designing containers on mobile and IoT devices. Morabito et al. implemented containers on Raspberry Pi 2 and evaluated the performance with sample benchmark problems [7]. They observed the performance of CPU, DISK I/O, and On-line Transaction Processing using sysbench benchmark problems. However, they found a very small improvement of 2.7% for CPU and 6 and 10% of memory and I/O improvement. Dupont et al. presented Cloud4IoT that offers migration during roaming or offloading of IoT functions [24]. Their proposed model automatically discover an alternative gateway and informs the cloud orchestrator by sending signals. The cloud orchestrator updates the Node affinity of its data processing container. They also proposed to find an optimal place of containers during offloading of IoT function so that less data requires to be transferred to the cloud. However, they did not discuss the process of finding the gateway or optimal place. Lee et al. analyzed the network performance for a container-based model running on IoT devices [25]. Their experimental result showed that a container-based system provides an almost similar level of network performance as provided by a native environment. All the above-mentioned works primarily showed the comparison of creating containers in IoT device with the traditional setup. In order to use containers in an IoT-based cloud platform, we need to have a model that will show the container formation, container orchestration, and migration. Therefore, In our model, we present the process of container creation, container orchestration, task scheduling and distribution, and container migration.

5 Conceptual Architecture

The conceptual architecture of IoTDoc is depicted in Fig. 2. The three primary components are Cloud Manager (CM), Swarm Manager (SM), and Node Manager (NM). CM is the central part of the cloud system that creates interaction between a user and an IoT provider. A new user needs to register through the management unit for receiving any service. Management unit also verifies the existing users and allow them to explore information associated with their accounts. Multiple users can request for service during a certain time interval. Design and Redesign unit analyze the existing resources and creates swarm managers. It also splits an existing swarm manager or merges or deletes a swarm manager. The scheduling and provisioning unit organizes the job received from multiple users and schedules them to a swarm manager. Monitoring unit observes the workload and performance of each swarm manager and removes or migrate the assigned job if necessary. The SM is the head of a cluster formed by a group of devices. An SM has three primary units; Scheduling and Allocation, Migration, and Reporting Unit. Scheduling and Allocation unit contacts the NMs, analyzes the resources and sends the image of a task and associated dependencies to NMs. In case an NM is overloaded, it informs the SM, and the migration unit in SM decides the task needs to be terminated or migrated. The reporting unit keeps track of the work done by an NM and sends the information to CM for generating the bill associated with a task. Additionally, it also collects any partial result depending upon the task type and sends that to a user via CM. Each devices has an NM that manages the containers running in the device. An NM has two primary units; Install/Uninstall and Allocation/Reallocation. Install/Uninstall unit generates or removes a container corresponding to a task. Allocation/Reallocation unit schedules and allocates resources for a specific container. It also reschedules or reallocates resources for the existing containers when a new container is installed or the existing container is deleted.

We utilized Docker's Swarm feature to orchestrate and manage our cluster of Docker engines. This feature can be used to create a "'swarm", or a cluster, of nodes, deploy application services to the swarm and manage the behavior of the swarm. SM deals with managerial tasks including maintaining the cluster state, scheduling services and serving swarm mode HTTP API endpoints [26]. SM use the Raft algorithm to maintain a consistent swarm state. NM needs at least SM to work properly and

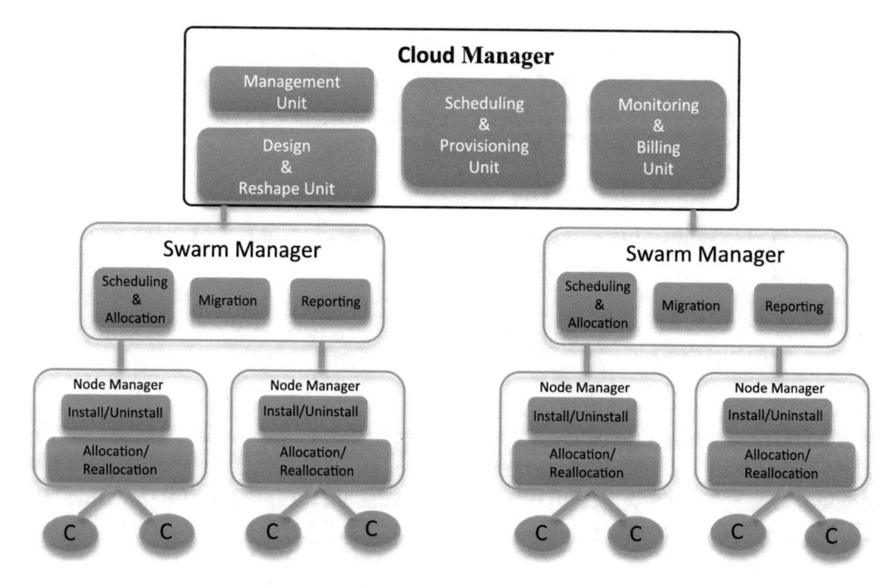

Fig. 2 Conceptual architecture of IoTDoc

their purpose is to execute tasks given to them by their SM. These tasks carry Docker containers and include containers and the commands necessary to run inside the container. SM uses internal load balancing to distribute these tasks to nodes within the cluster.

6 Operational Model

In this section, we discuss the process of cluster formation and Swarm Manager (SM) selection, resource allocation, distribution, and installation of container images, and container migration.

6.1 Swarm Manager (SM) Selection

We use weighted clustering algorithm proposed by Chatterjee et al. [27] for SM selection. The algorithm assigns a weight to each node based on the ideal node degree, transmission power, mobility, and battery power. Their algorithm also provides flexibility in adjusting the weighting factors, unlike many other algorithms, making it a preferred algorithm for possible extensions of our experiment. It is an iterative process beginning with the identification of the neighbors of each node v,

$$d_v = |N[v]| = \bigcup_{v' \in V, v \neq v} \{v' | dist(v, v') < tx_{range}\}$$

where tx_{range} is the transmission range of v. Then the degree-difference, $\Delta_v = |d_v - \delta|$, is computed. The sum of distances D_v of each node is calculated by,

$$D_v = \sum_{v' \in N(v)} \{dist(v, v')\}$$

The running average of the speed for every node until current time T gives a measure of mobility,

$$M_v = \frac{1}{T} \sum_{t=1}^{T} \sqrt{(X_t - X_{t-1})^2 + (Y_t - Y_{t-1})^2}$$

where (X_t, Y_t) and (X_{t-1}, Y_{t-1}) are the coordinates of the node v at time t and $(t-1)$, respectively. Then the cumulative time, P_v, in which a node v acts as a SM is calculated. All of these intermediate calculations factor into the combined weight W_v for each node v in which

$$W_v = w_1 \Delta_v + w_2 D_v + w_3 M_v + w_4 P_v$$

and w_1, w_2, w_3, and w_4 are weighting factors, allowing the node selection process to be adaptable. The node with the smallest weight W_v is selected as the SM. The steps above are repeated for remaining nodes that are not an SM, or a neighbor of current SM. Each node monitors the signal strength it receives from the SM and notifies the SM if its signal is too weak. The node will be attempted to be transferred to another nearby cluster that have an SM. If there is no available cluster to adopt the node, the SM selection algorithm is invoked and the network is recreated.

The cardinality of a SM's cluster size represents the SM's load and the variance of the cardinalities signifies the distribution of the load. For distributing the load efficiently, we introduce a term load balancing factor (LBF) that is the inverse of the variance of the cardinality of the clusters as defined by the following equation,

$$LBF = \frac{n_c}{\sum_i (x_i - \mu)^2} \text{ and } \mu = \frac{(N - n_c)}{n_c}$$

Here, n_c is the number of SMs, x_i is the cardinality of cluster i, and μ is the average number of neighbors of an SM where N is the total number of nodes in the system. Higher LBF values represent better load distributions. As the LBF tends towards infinity, the system tends towards perfect balance.

6.2 Resource Allocation

We have a pool of IoT devices where we need to deploy the containers. We initially consider the processing power and memory size of the IoT devices during scheduling the containers. We estimate the CPU and memory utilization of individual contain-

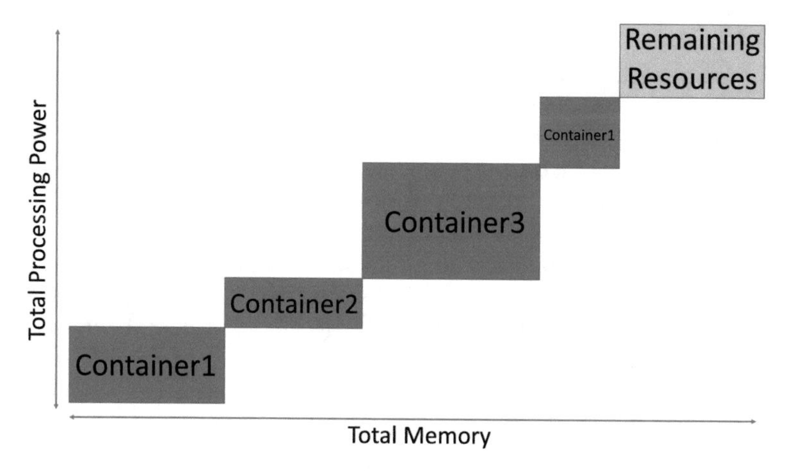

Fig. 3 CPU and memory allocation for containers

ers and add them together to determine the total CPU and memory utilization. For example, if we have two containers, where the first container requests 30% CPU and 25% memory whereas the second container requests 25% CPU and 20% memory then the total request will be the vector sum of the first and second request [28].

However, the resource is not limited to only CPU and memory. There might be several other types of resources a container might need, such as storage, sensors etc. Therefore, each container's size is represented by an m-dimensional vector where each dimension is associated with a resource type. An example scenario considering just CPU and memory is shown in Fig. 3 where we consider the deployment of three containers in a single IoT device. We intend to reduce resource wastage during container placement. We used the model presented by Gao et al. [28] to minimize resource wastage.

$$W_t = \frac{|R_t^c - R_t^m| + \delta}{U_t^c - U_t^m}$$

W_t represents the resource wastage of tth IoT device. R^c and R^m represents normalized remaining CPU and memory resource. U^c and U^m is the normalized CPU and memory utilization. We add a small δ value of approximately 0.000001 for balancing the remaining resources efficiently in all dimensions.

We adopt particle swarm optimization (PSO) method that is a stochastic, population based meta heuristic model [29, 30] for placing the containers on the IoT devices. PSO uses a two-dimensional encoding scheme where the first is an n-bit vector $S^t i1, S^t i2, \ldots S^t in$ represents the n number IoT devices and the second dimension represents a subset of containers that will be placed on those IoT devices [31]. For example if a bit in the first dimension is set 1, the corresponding IoT device is active and at least one container is running on that IoT device. On the other hand, a 0 indicates that the corresponding IoT device is idle and unused. We use another n dimensional vector $P^t i1, P^t i2, \ldots P^t in$ knows as particle velocity. A 0 in kth position of the velocity vector indicates that the containers running on kth IoT device must be reevaluated whereas 1 indicates we do not need to do any change in the device at this point.

The subtraction operation represented by Θ indicates the difference between two the proposed container placement solutions. For example, if the first solution $S_i = (1, 0, 1, 1)$ and the second solution $S_j = (1, 0, 0, 1)$ then $S_i \Theta S_j = (1, 1, 0, 1)$.

The addition operation $\oplus$ is used to update the particle velocity. We consider three factors, current velocity inertia F_1, local best position F_2 and global best position F_3. For example, if we have two solution first with velocity $VS_i = (0, 1, 0, 1)$ with probability $P_i = 0.4$ and the second velocity $VS_j = (1, 1, 0, 0)$ with probability $P_j = 0.6$ then $VS_i \oplus VS_j = (?, 1, 0, ?)$. Therefore, we can say that we need to change second bit while the third bit should remain unchanged. The ? sign indicates uncertainty and can be computed from F_1, F_2, and F_3 as follows [31].

$$F_{1i} = \frac{1/f(S_i^t)}{1/f(S_i^t) + 1/f(S_{locBest}^t) + 1/f((S_{globBest}^t)}$$

$$F_{2i} = \frac{1/f(S_{locBest}^t)}{1/f(S_i^t) + 1/f(S_{locBest}^t) + 1/f((S_{globBest}^t)}$$

$$F_{3i} = \frac{1/f(S_{globBest}^t)}{1/f(S_i^t) + 1/f(S_{locBest}^t) + 1/f((S_{globBest}^t)}$$

Here $f(X_t)$ denotes the fitness of ith particle solution and is directly proportional to the resource wastage. Therefore, large fitness value for current solution indicates more wastage and less probability of selecting the current solution. For a IoT device i, the current fitness is the sum of CPU and memory utilization of all the containers running on that device. i.e.

$$f(S_j^t) = \frac{\sum_{i=1}^{n} \lambda_{ij} R_i^p}{T_{current_j}^p} + \frac{\sum_{i=1}^{n} \lambda_{ij} R_i^m}{T_{current_j}^m}$$

Here $T_{current_j}^p$ and $T_{current_j}^m$ indicates the maximum CPU and RAM (threshold value) allocated in jth IoT devices and R_i^p and R_i^m are the CPU and RAM require for i^{it} container. λ_{ij} is a binary value where 1 indicates ith container is allocated on jth IoT device, otherwise 0. We use a similar fitness function for local best and global best.

$$f(S_{locBest}^t) = \frac{\sum_{i=1}^{n} \lambda_{ij} R_i^p}{T_{locBest_j}^p} + \frac{\sum_{i=1}^{n} \lambda_{ij} R_i^m}{T_{locBest_j}^m}$$

$$f(S_{globBest}^t) = \frac{\sum_{i=1}^{n} \lambda_{ij} R_i^p}{T_{globBest_j}^p} + \frac{\sum_{i=1}^{n} \lambda_{ij} R_i^m}{T_{globBest_j}^m}$$

We generate a random value r between 0 and 1 and set the value of uncertain bit b as follows:

b = b1, bit value before updating if $r <= F_{1i}$

b = b2, bit value of particle local best velocity if $F_{1i} < r < F_{2i}$

b = b3, bit value of particle global best velocity if $F_{2i} < r < F_{3i}$.

For this local position, local fitness is defined as the sum of CPU and memory utilization of all the VMs that are running on that local server. The local fitness of jth server can be calculated by equation given below.

The multiplication operation denoted by $\times$ is used for updating the particle. For example, if a particle position $S_i = (1, 0, 1, 1)$ and particle velocity $V_i = (1, 1, 0, 0)$ then $S_i \otimes V_i = (1, 0, 0, 0)$.

6.3 Image Distribution and Installation

In IoTDoc, each user provides a summary of the tasks that they are planning to perform. The analyzer performs thorough analysis and figures out the common system libraries users are planning to use. The analyzer groups together all the libraries and creates a base image. Additionally, if a large number of users are also planning to use some application libraries, it can list those libraries into the base image and expand the base image. This base image can be considered a stack of read-only layers each associated with a command executed as specified by the analyzer [32]. The process of image distribution and container formation is depicted in Fig. 4.

The Cloud Manager analyzes the resources and distributes the images among the Swarm Managers (SM). Each SM creates a certain number of files known as docfile

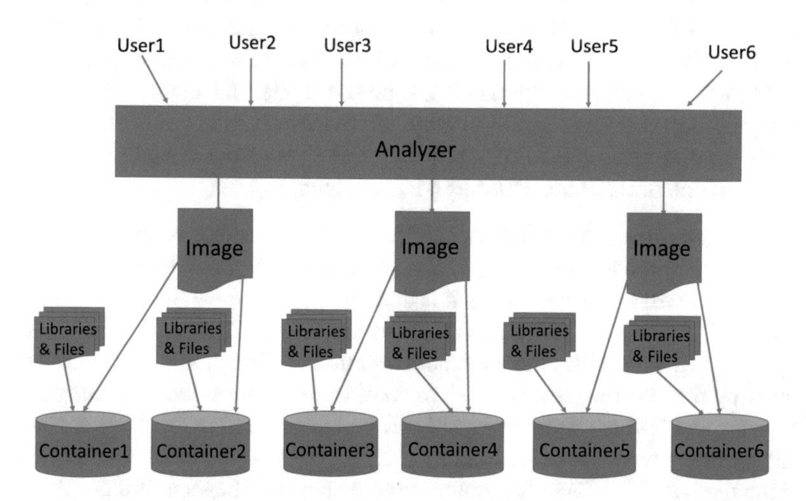

Fig. 4 Process of image distribution and container formation

each associated with a user task. The docfile specifies the list of dependencies along with the command needs to use for installing those dependencies associated with an application. The SM verifies the resources of individual IoT devices and performs a mapping of image and Node Manager (NM) where each NM receives one base image and installs the image. Next, the Swarm manager creates a separate docfile for each user that contains a set of instructions users need to execute in order to run their respective applications. Based on the base image requires for executing the instructions in a docfile, it is scheduled to a specific NM. The NM creates a container by executing the set of instructions written in docfile in base image. The execution of each command in the docfile updates the current status of the existing container and creates an additional layer on top of its previous layer.

6.4 Migration

In an edge computing environment, migration can happen either when the device moves to a different area under a different SM or when it has any power, hardware, or software failure. Since the location can be changed, we follow a very similar approach as proposed by Ma et al. [33]. We consider the two possible scenarios during migration.

A Node Manger Fails: The corresponding SM retrieves the last update from the summarized information associated with the NM. It creates an image group consists of the images of all the docker containers of the NM from this update. Cloud manager scans all the other available NMs under its coverage area, selects an NM of similar configuration, and installs the image to the selected NM. Otherwise, it sends the image to the Cloud Manager (CM). CM queries all the other SMs to find out the availability of any NMs. CM waits a threshold time to receive any response of available NMs from all the SMs. If an SM replies within this threshold time, the image is sent to it. The SM installs it to that matching NM. If no SM replies, the cloud manager breaks the image groups into multiple subgroups of images. It again starts querying to all the SMs for a suitable match for each image and sends the image to that SM for installing to one of the NMs under its coverage area.

Node Manger Moves and Want to Continue: When a device is moving it can still perform the required job. However, maintaining the synchronization or keeping track of the updates will be difficult since it is done by its respective SM. Therefore, first, the current SM sends the summarized information to the CM. A threshold time t_1 is set by the CM that specifies the maximum duration the SM should wait before either considering the NM for the same task or continue to perform the remaining portion of the task. The NM can only be considered for the continuing for the task if it joins an SM within t_1 time. However, joining an SM does not guarantee that an NM can continue the task. The CM sets another time t_2 that specifies the minimum amount of time it has to be connected with that SM. If the NM agrees the time t_2 then it can continue the task and the corresponding SM receives the summarized information from the CM. The CM copies the layers associated with each container in the NM, summarizes those layers, and compares with the summary that it has received from the CM. If the summary matches it sends an acknowledgment message to the NM and the NM continues its assigned task.

Therefore, first, we anticipate the device's future location and the nearest SM from that location. If that future SM does not contain the base layer of any container that performs the task, then it starts synchronizing the base image of the container with the current SM.

The SM requests the NM to generate a snapshot that contains the runtime memory information of containers. While the containers inside the device are still running, the current SM finds the predicted future SM. A migration request message is sent from the current SM to the future SM. The SM sends a suspension message to the NM and the NM suspend all the operations performed in a container. In the meantime, the future SM receives the base layer information from the current SM and select NMs for creating container using the newly created base layer.

7 Experimental Evaluation

We designed a prototype of IoTDoc using sample Raspberry Pis. We considered benchmarks tasks provided in Sysbench repository and created containers accordingly. Sysbench is a suite that provides a quick and easy platform to get information about a system or testbed and see how it would run under an intensive load. In our sample design, we did not consider node failures and therefore, we assumed that we were not required to perform container migration. We designed a similar distributed model by considering virtual machines in Amazon Web Service (AWS) and ran the Sysbench problems. We compared the CPU, Data Definition (DD) read/write, and memory performance of both Swarm Managers (SM) and Node Managers (NM) running in IoTDoc with the VM-based system and the result is depicted in Figs. 5,

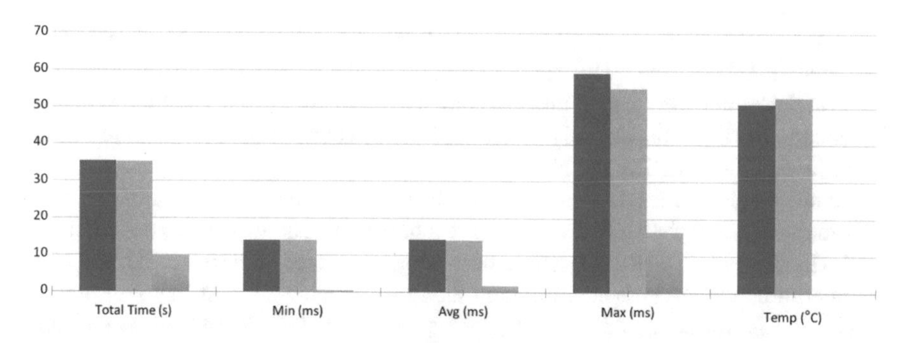

Fig. 5 CPU test performance comparison between Amazon EC2 and IoTDoc

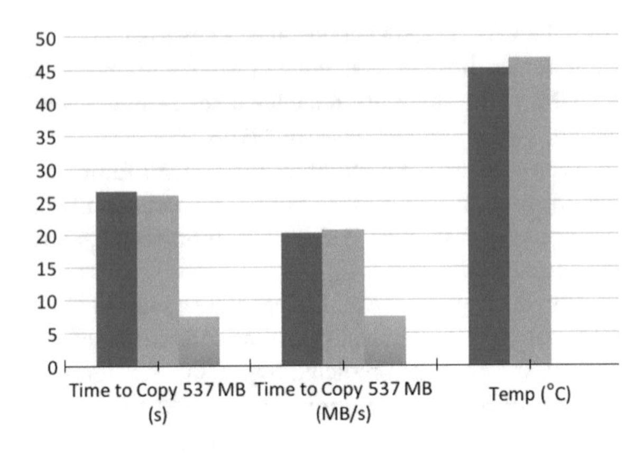

Fig. 6 DD read test performance comparison between Amazon EC2 and IoTDoc

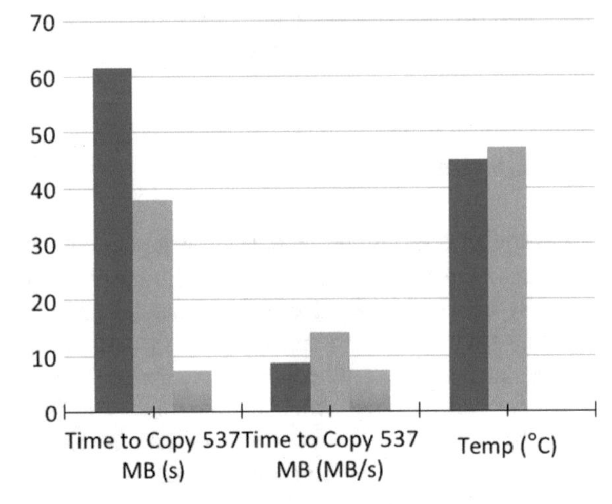

Fig. 7 DD write test performance comparison between Amazon EC2 and IoTDoc

6, 7, 8, 9, 10. Since we do not have direct access of the EC2 physical machines, we could not compare the heat. We just showed the heat generated in our SM and NM.

Figure 5 shows the results of the SM, NM, and AWS after a CPU test. This particular test checks the performance of the CPU and uses a maximum prime number value to determine how long it takes to calculate all the numbers. The EC2-t2-micro instance performed substantially better than the IoTDoc, with approximately three times faster than both SM and NM.

Figure 6 shows the DD Read test which measures how fast a system can read information. A Raspberry Pi does not have a hard drive and instead uses a microSD card. Therefore, the DD Read test measures the performance of the microSD card. The test measures the RAM availability and then manipulates a file of that size, which reveals the performance of the disk processing speed. The EC2-t2-micro instance outperforms both SM and NM.

The DD Write test shown in Fig. 7 measures how fast a system can write information. This test is similar to the DD Read test, measuring the performance of the

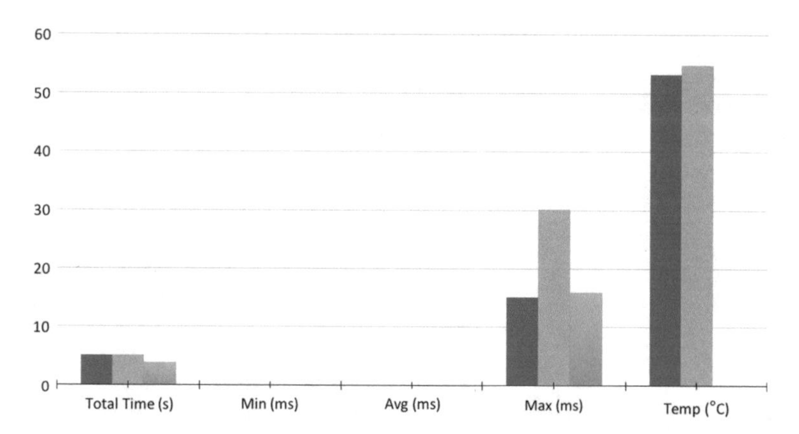

Fig. 8 Memory test performance comparison between Amazon EC2 and IoTDoc

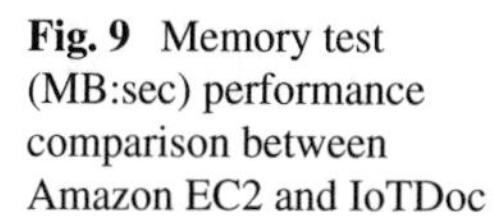

Fig. 9 Memory test (MB:sec) performance comparison between Amazon EC2 and IoTDoc

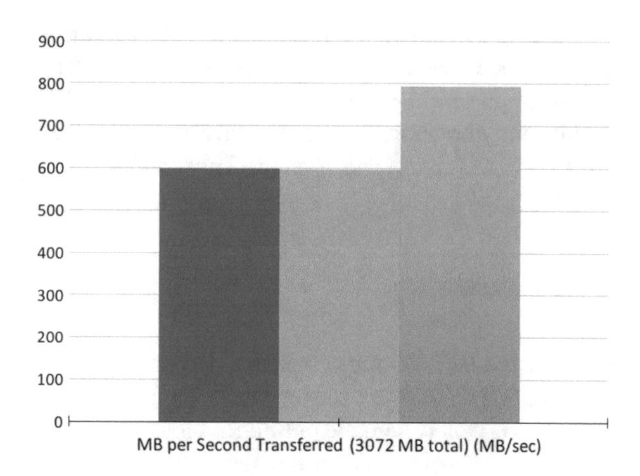

microSD card. However, unlike DD read test, the result appears to be a much greater difference in the time to copy between the SM and NM in case of DD write test. There also appears to be less of a difference between the performance of the IoTDoc and the EC2-t2-micro instance with the Time to Copy (MB/s) measurement. This test provides positive results for the case of using IoTDoc over a more expensive service through AWS.

The Memory test in Fig. 8 measures the overall health of the RAM, and the average amount of available RAM. The IoTDoc performs similarly to the EC2-t2-micro instance, with the SM even performing slightly better than the instance in the Max Time category.

Figure 9 displays a more specific type of Memory test involving the data transfer speed. It calculates the performance of the RAM by measuring the speed of transferring data, in MB per second. The EC2-t2-micro instance performs better than the IoTDoc but not as dramatically as in the CPU and DD Read tests.

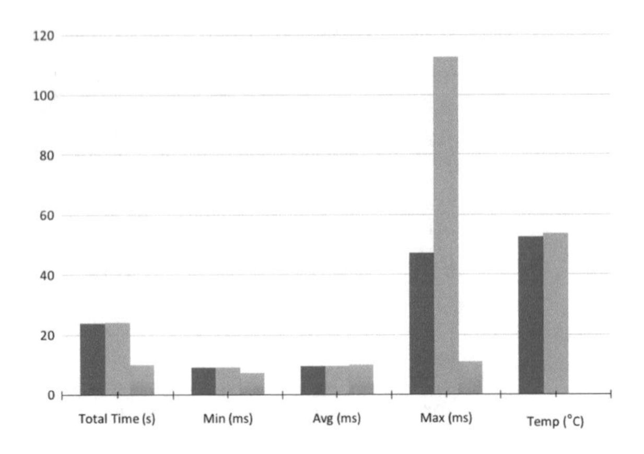

Fig. 10 Threads test performance comparison between Amazon EC2 and IoTDoc

Finally, the Threads test shown in Fig. 10 is a test that benchmarks scheduler performance. Each thread runs requests and locks until the task is finished. The thread is placed in a queue by a scheduler. The EC2-t2-micro instance performs significantly better in Total Time and especially Max Time, but is comparable to the IoTDoc in Minimum and Average Time, which is really important.

8 Discussion

Overall, the EC2-t2-micro instance outperforms the IoTDoc in most of the benchmarking tests. However, in some tests, the EC2 and IoTDoc is comparable, showing that the IoTDoc is a viable option for cloud computing. Also, with a one-time price of $35 for a Raspberry Pi compared to the accruing costs of utilizing AWS (around $100 a year based on usage), the IoTDoc provides a more affordable, cost-effective alternative to large platform cloud computing services. This also makes IoTDoc an excellent option as a learning platform, providing sufficient computing power at an accessible cost. However, our motivation to present IoTDoc is not to develop a system that outperforms VM in terms of computation. Instead, we would like to present a cost-efficient system that will be easy to install, manage, migrate. We also intend to use the system in low-cost IoT-enabled cloud platform and perform a wide range of tasks that the normal VM-based systems fail to perform. Our experimental data showed that IoTDoc can be used for solving computational intensive problems. We can expect that with technological advancements, IoT devices will be more powerful and therefore, the performance gap between IoTDoc and VM-based system will be reduced in near future.

9 Conclusion

Though hardware-level virtualization is in growing demand, its deployment is time-consuming and expensive as compared to OS-level virtualization. Moreover, it is difficult to install such a virtualization platform in a resource-constrained IoT device. Therefore, in this paper, we presented IoTDoc, an OS-level virtualization platform that deploy containers on IoT based distributed system. In our operational model, we discussed the possible challenges along with strategies for swarm manager, scheduling, image distribution, and migration. In order to show the effectiveness of our proposed system, we implemented a prototype of IoTDoc and ran some sample benchmark tasks as a proof of concept. Our result showed that IoTDoc can be used as an effective alternative of a VM-based system in near future.

References

1. Roundup of cloud computing forecasts and market estimates, 2018. https://www.forbes.com/sites/louiscolumbus/2018/09/23/roundup-of-cloud-computing-forecasts-and-market-estimates-2018/#5e638a7507b0 (2018)
2. Noor, S., Hasan, R., Haque, M.: Cellcloud: a novel cost effective formation of mobile cloud based on bidding incentives. IEEE Cloud (2014). https://doi.org/10.1109/CLOUD.2014.36
3. Noor, S., Hossain, M., Hasan. R.: Sascloud: ad hoc cloud as secure storage. In: IEEE BDCloud (2016)
4. Tihfon, G., Park, S., Kim, J., Kim, Y.: An efficient multi-task paas cloud infrastructure based on docker and aws ecs for application deployment. Clust, Comput (2016)
5. Naik, N.: Building a virtual system of systems using docker swarm in multiple clouds. In: IEEE ISSE (2016)
6. Bellavista P., Zanni, A.: Feasibility of fog computing deployment based on docker containerization over raspberrypi. In: IICDCN (2017)
7. Morabito, R.: A performance evaluation of container technologies on internet of things devices. In: IEEE INFOCOM Workshop (2016)
8. Celesti, A., Mulfari, D., Fazio, M., Villari, M., Puliafito, A.: Exploring container virtualization in iot clouds. In: IEEE SMARTCOMP (2016)
9. Mulfari, D., Fazio, M., Celesti, A., Villari, M., Puliafito, A.: Design of an IoT cloud system for container virtualization on smart objects. In: ESOCC (2016)
10. Renner, T., Meldau, M., Kliem A.: Towards container-based resource management for the internet of things. In: ICSN (2016)
11. What is kubernetes? https://www.redhat.com/en/topics/containers/what-is-kubernetes
12. Wheatley, M.: New working group aims to bring kubernetes to IoT edge networks. https://siliconangle.com/2018/09/26/new-working-group-created-bring-kubernetes-iot-edge-networks/ (2018)
13. Christopher, B.: A fast, new way to spot problems in kubernetes. https://developer.ibm.com/blogs/2018/11/01/spot-problems-in-kubernetes-quickly/
14. Mobile cloud computing solutions, architecture, app, analytics. https://www.v2soft.com/services/technology/mobility-solutions/cloud-mobility-solutions (2018)
15. What is a container. https://www.docker.com/what-container (2018)
16. Vaughan-Nichols, S.: What is docker and why is it so darn popular? https://www.zdnet.com/article/what-is-docker-and-why-is-it-so-darn-popular/
17. What is docker? https://docs.microsoft.com/en-us/dotnet/standard/microservices-architecture/container-docker-introduction/docker-defined (2018)

18. Columbus, L.: 2018 roundup of internet of things forecasts and market estimates. https://www.forbes.com/sites/louiscolumbus/2018/12/13/2018-roundup-of-internet-of-things-forecasts-and-market-estimates/#259c9b707d83 (2018)
19. Halladay, J.: Moore's law and the future of IoT. https://medium.com/mybit-dapp/moores-law-and-the-future-of-iot-d9ed7d725f0a (2018)
20. Ha, K., Abe, Y., Chen, Z., Hu, W., Pillai, P., Satyanarayanan, M.: Adaptive vm handoff across cloudlets, Brandon Amos (2015)
21. Felter, W., Ferreira, A., Rajamony, R., Rubio, J.: An updated performance comparison of virtual machines and linux containers. In: International Symposium on Performance Analysis of Systems and Software (ISPASS) (2015). https://doi.org/10.1109/ISPASS.2015.7095802
22. Agarwal, K., Jain, B., Porter, D.E.: Containing the hype. In: APSys (2015)
23. Sharma, P., Chaufournier, L., Shenoy, P., Tay, Y.C.: Containers and virtual machines at scale: a comparative study. In: International Middleware Conference (2016)
24. Dupont, C., Giaffreda, R., Capra, L.: Edge computing in IoT context: horizontal and vertical linux container migration. In: GIoTS (2017)
25. Lee, K., Kim, H., Kim, B., Yoo, C.: Analysis on network performance of container virtualization on IoT devices. In: ICTC (2017)
26. Docker swarm 101. https://www.aquasec.com/wiki/display/containers/Docker+Swarm+101 (2018)
27. Chatterjee, M., Das, S.K., Turgut, D.: Wca: a weighted clustering algorithm for mobile ad hoc networks. Clust. Comput. **5**, (2002)
28. Gao, Y., Guan, H., Qi, Z., Hou, Y., Liu, L.: A multi-objective ant colony system algorithm for virtual machine placement in cloud computing. J. Comput. Syst. Sci. **79**, (2013)
29. Patnaik, S.S., Panda, A.K.: Particle swarm optimization and bacterial foraging optimization techniques for optimal current harmonic mitigation by employing active power filter. Appl. Comp. Intell. Soft Comput. (2012)
30. Parsopoulos, K.E., Vrahatis, M.N.: Particle Swarm Optimization and Intelligence: Advances and Applications. IGI Publishing, Information Science Reference-Imprint (2010)
31. Wang, S., Liu, Z., Zheng, Z., Sun, Q., Yang, F.: Particle swarm optimization for energy-aware virtual machine placement optimization in virtualized data centers. In: 2013 International Conference on Parallel and Distributed Systems (2013)
32. Docker inc. 2017. docker images and containers. https://docs.docker.com/engine/userguide/storagedriver/ (2017)
33. Ma, L., Yi, S., Li, O.: Efficient service handoff across edge servers via docker container migration. In: ACM/IEEE SEC (2017)

A Survival Analysis-Based Prioritization of Code Checker Warning: A Case Study Using PMD

Hirohisa Aman, Sousuke Amasaki, Tomoyuki Yokogawa and Minoru Kawahara

Abstract Static code analysis tools (code checkers) scan source programs and issue warnings to potentially-problematic parts. Programmers can utilize a code checker whenever they change their source code to make sure that their code changes do not carry high risks of decreasing the code quality. Although code checkers would be helpful to detect risky code changes as early as possible, there is a practical problem which prevents an active utilization of such tools in the real: code checkers tend to produce a lot of false-positive warnings, i.e., such a tool outputs many warnings, but the majority of them are not attractive to the programmer. Toward an efficient utilization of code checkers, this paper proposes an application of the survival analysis method to prioritize code checker warnings. The proposed method estimates a warning's lifetime with using the real trend of warnings through code changes; the brevity of warning means its importance because severe warnings are related to problematic parts which programmers would fix sooner. This paper conducts a large-scale case study of 6,927,432 warnings (259 types of warnings) appeared in 100 open source software projects. The results show that only 30 types of warnings are practically important for programmers in terms of the brevity, and the proposed method can drastically reduce the number of really needed warnings.

Keywords Code checker warning · Survival analysis · Warning prioritization

H. Aman (✉) · M. Kawahara
Center for Information Technology, Ehime University, Matsuyama, Ehime, Japan
e-mail: aman@ehime-u.ac.jp

M. Kawahara
e-mail: kawahara@ehime-u.ac.jp

S. Amasaki · T. Yokogawa
Faculty of Computer Science and Systems Engineering, Okayama Prefectural University, Soja, Okayama, Japan
e-mail: amasaki@cse.oka-pu.ac.jp

T. Yokogawa
e-mail: t-yokoga@cse.oka-pu.ac.jp

R. Lee (ed.), *Big Data, Cloud Computing, and Data Science Engineering*,
Studies in Computational Intelligence 844,
https://doi.org/10.1007/978-3-030-24405-7_5

1 Introduction

Software products have usually evolved/upgraded through many function enhancements and fault fixes. While a source code change is inevitable in software evolution and upgrade, a code change also has risks of introducing a new fault or deteriorating the code quality such as the understandability and the maintainability in practice [1–3]. Furthermore, an immature code or a poor quality code may be released through code changes due to the shortage of time to a release of an updated version—it is referred to as a technical debt [4–6].

During the software development and maintenance, it is a significant challenge to avoid the above risks such as making new faults, decreasing the code quality or producing technical debts, for successful quality management. The ideal activity is to thoroughly review all source programs whenever a programmer changed source code. However, it is hard to frequently perform such careful code reviews for a large-scale and complex software product since a manual code review by an expert is a high-cost activity. To support successful and efficient code reviews, static code analysis tools (code checkers) have been developed (e.g., PMD,[1] SpotBugs[2]). These tools can automatically scan all source programs by their predefined rules, and detect potentially problematic code fragments which may cause faults or may be poor quality parts having discouraged coding patterns or code smells. By utilizing such tools during the programming, we can expect an early detection of a risky code change to prevent deteriorating the code quality and introducing technical debts.

However, there has also been an obstacle to the active utilization of code checkers because such tools tend to produce many warnings, but the majority of them are false-positive, i.e., programmers did not need to fix the warned parts in many cases [7, 8]. That is one of the main reasons why programmers do not actively use such tools [9–11]. Thus, a proper prioritization of warnings has been a challenge for efficient utilization of code checkers. Although many rule sets of code checkers assign (predefined) priorities to warnings (e.g., PMD uses five-grade priority), most of the warnings appearing in real source programs are likely to be the "middle level" [12], so it would not be practical to use the default priorities. To obtain a proper set of priorities, there have been studies which utilize feedback from tool users in the past [13, 14]. Although getting feedback from users is a useful way of evaluating warnings, there is also another kind of problem: the feedback collection effort. Since many warnings may appear in large-scale software, tool users would have to put in much effort on providing feedback.

To solve the above problem of feedback collection effort, we propose another method for automatically deciding a proper priority of warning, which focuses on the changes of warnings through commits to the repository, and does not need to get tool users involved in the feedback collection. Our proposal is an application of the survival analysis [15]: we trace each warning from its birth to death through upgrades of a source file—its birth means the warning is started appearing in a source file,

[1] https://pmd.github.io/.

[2] https://spotbugs.github.io/.

and its death means the warning disappears—, and estimate the lifetime by using the survival analysis method. A shorter-life warning would be more important since many programmers tend to resolve it sooner. On the other hand, a longer-life warning would be lower-valued because many programmers have left it for a longer time. Thus, a warning having a shorter lifetime should have a higher priority to be reviewed, and a code change which produces such a short-life warning may be risky in terms of the code quality.

The main contributions of this paper are as follows:

(1) We propose a novel method for prioritizing code checker warnings, which does not need to ask users to provide their feedback. The proposed method utilizes the survival analysis method.
(2) We report large-scale case study of the survival analysis on code checker warnings: 6,927,432 warnings (259 kinds of warnings) appeared in one or more versions of 54,162 source files from 100 open source software (OSS) projects.

The structure of the paper is as follows. Section 2 describes the related work. In Sect. 3, the code checker warnings and their lifetimes are explained, and the survival analysis-based evaluation method is proposed. Section 4 reports on our case study and gives discussions about the results. Finally, Sect. 5 presents our conclusion and future work.

2 Related Work

There have been studied to obtain a proper set of priorities of code checker warnings. Shen et al. [13] proposed a method to customize the priorities of warnings by using feedback from the users and implemented it as an extension of FindBugs.[3] They performed customization of the warning priority predefined in FindBugs through analyzing the source files of the Java Developer Kit, where four students provided feedback to the warnings. After that, they conducted an assessment experiment with three OSS projects, then reported their tool had a lower false-positive rate than the original FindBugs.

Sadowski et al. [14] developed the program analysis ecosystem, TRICODER, in which the programmers can evaluate the warnings, and TRICODER automatically tunes the priority of warning by learning the feedback. TRICODER can contain two or more code analyzers and can show not only the warnings but also possible solutions of them. Tool users click links shown in TRICODER GUI, which correspond to feedback, e.g., "not useful" link signifies that the warning was not needed. By collecting such feedback data and computing the "not-useful" rate of warning, TRICODER produces a better priority set of warnings. They introduced the tool into Google's development environment and showed its usefulness.

[3] FindBugs was one of the popular tools, but it is not supported now. SpotBugs is the successor of FindBugs.

While the methods proposed by Shen et al. and Sadowski et al. are remarkable solutions to the warning prioritization problem, they need to get programmers involved in the tool turning. When a large-scale project newly started to use such a tool, the programmers would be required much effort for answering many questions from the tool. As yet another solution of the prioritization problem, we evaluate warnings by focusing on the code repository in this paper, where we do not need to ask users to provide feedback. This idea is based on the study by Burhandenny et al. [12]; Burhandenny et al. proposed to evaluate warnings by using the changes in the number of warned parts through upgrades, where an increment of warning makes the priority lower and a decrement does the priority higher. Since their work did not take into account the duration of time between changes, we perform a further study in this paper by applying the survival analysis method.

3 Code Checker Warning and Survival Analysis

3.1 Code Checker Warning

As mentioned above, a code checker is a tool for scanning source programs under predefined rules. By using such a tool, programmers can quickly check their programs. Moreover, a project manager can see whether an upgrade does not introduce any new risk to the product or not, by comparing the scanning results before and after the code change.

As an example, we scanned 1,632 Java source files (273 KLOC) included in RxJava project, by using PMD 6.5.0 with all Java-related rule sets[4]; PMD is one of the popular checkers. The scanning took about 86 s (0.05 s per source file) on our environment[5] and produced 190,617 warnings (117 warnings per source file). Table 1 shows the number of the warnings by the predefined priority where the larger value signifies the higher priority. As pointed out in the previous work [8, 12], many warnings appeared, and the majority of them (91.4%) were the middle-level (priority 3) ones. Since the project has been successfully released their products for years, it seems to be quite unlikely that all of the warnings are severe for quality management. In other words, many of these warnings may not be useful even though their predefined priorities are not low. If we successfully singled out important warnings buried in a heap of disregarded ones, developers and project managers could easily detect risky code changes as early as possible.

Now the challenge we face is how to evaluate the importance of a warning automatically. As we mentioned above, it does not look a good way to use the default (predefined) priority. Although customizing the priority by using the feedback from tool users is a promising way, it may not be the best because we have to ask users

[4]https://github.com/pmd/pmd/blob/master/pmd-core/src/main/resources/rulesets/internal/all-java.xml.

[5]OS: Linux 3.10.0, Java: OpenJDK 1.8.0_201, CPU: Intel Core i5 3.2GHz, Memory: 16GB.

Table 1 Number of warnings by the predefined priority

Priority	Number of Warnings	(Percentage %)
1	608	(0.3)
2	1,392	(0.7)
3	174,262	(91.4)
4	4,343	(2.3)
5	10,012	(5.3)

to have much effort for the customization. Thus, in this paper, we propose to focus on the changes of warnings throughout commits to the code repository (code change events). By using the code repository, we can observe the practices which are common to many programmers and evaluate the importance of warnings without directly asking for feedback from the tool users.

3.2 Changes of Code Analyzer Warnings Through Commits to Code Repository and Warnings' Lifetimes

When a source file has been managed with a version control system such as Git, the system has tracked all code changes performed to the source file on its code repository. Hence, we can easily obtain the source files before and after a code change (a commit to the repository). Then, we can also observe the warning changes through the code change by running a code checker for both the older version of the source file and the newer version of it, and comparing their warnings.

Now we define the birth of warning, the death of warning, and the lifetime of warning below.

Definition 1 (*Birth of warning*) For a warning which appears in a source file, we define the birth of the warning as the event that the warning started to appear in the source file. A birth of warning occurs when we create a new source file or modify an existing source file. □

Definition 2 (*Death of warning*) For a warning which had appeared in a source file, we define the death of the warning as the event that the warning gets disappeared in the source file. A death of warning occurs when we modify or remove an existing source file. □

Definition 3 (*Lifetime of warning*) For a warning which already died, we define the lifetime of the warning as the time between its birth and its death. □

Figure 1 presents a simple example of three warnings w_1, w_2, and w_3 which appear in a source file. In this example, there have been four commits, i.e., four source code change events which occurred at t_1, t_2, t_3, and t_4, respectively. Warnings w_1, w_2, and

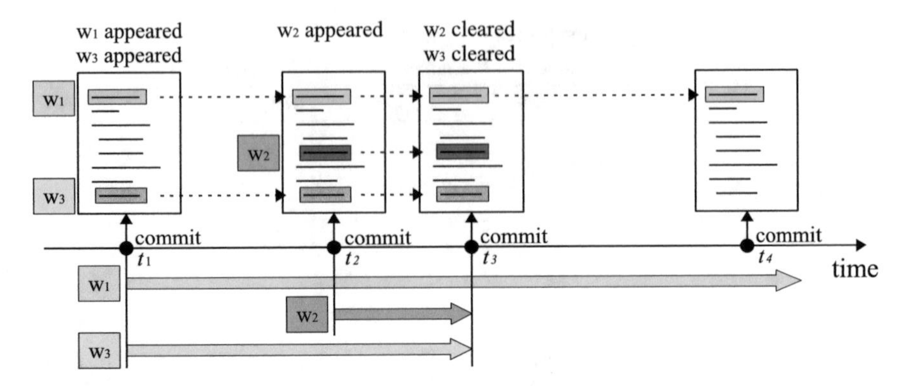

Fig. 1 An example of warnings (w_1, w_2, w_3) changes through commits

w_3 were born at t_1, t_2, and t_1, respectively. After that, two warnings w_2 and w_3 died at t_3. Thus, the lifetime of w_2 is $t_3 - t_2$, and that of w_3 is $t_3 - t_1$. Since w_1 is still alive, we cannot know its real lifetime; we statistically estimate the lifetime of a warning like w_1, i.e., a warning which has been alive at our observation, by the statistical method that we explain in the following Sect. 3.3.

Now we focus on the lifetime of a warning. If a warning has a longer lifetime, the warning had continued to appear for a longer time. In other words, a code fragment corresponding to the warning had remained unchanged for a longer time. We consider a reason why the programmer did not clear the warning. Regarding the use of code checker, there are two possible cases: (1) the programmer used a code checker and noticed the warning; (2) the programmer did not use any code checker, so he/she did not notice the warning.

(1) *The programmer used a code checker and noticed the warning*: If the programmer used a code checker and noticed the warning, he/she would have disregarded the warning. The programmer might consider that it is not a severe problem to be fixed, or it does not fit his/her programming style or practice.
(2) *The programmer did not use any code checker*: If the programmer did not use any code checker, he/she did not notice the warning. Nonetheless, the warning may reflect the programmer's style or practice. Since the programmer had kept the code fragment unchanged for a longer time, he/she may not acknowledge it as a problematic part to be fixed.

Hence, in both cases, a long-life warning is likely to be less important for the programmer.

On the other hand, we also consider a warning which has a shorter lifetime. A shorter lifetime of a warning represents that the programmer fixed the code fragment corresponding to the warning sooner after its birth, and moreover the warning was cleared by the fix. Thus, such a short-life warning is related to a code fragment which is likely to be fixed regardless of whether the programmer uses a code checker or not.

Therefore, we define the importance of warning by using its lifetime below.

Definition 4 (*Importance of warning*) For a warning made by a code checker, we define the importance of the warning as the brevity of its lifetime. □

To evaluate the importance of warning in practice, we need to properly deal with the following two issues (i) and (ii).

(i) *Same type warnings which have different lifetimes*: There are two or more warnings which are the same type of warning but have different lifetimes. For example, we faced five warnings which are referred to as "CommentSize" in one source file `FlowableRepeatWhen.java`[6] when we tried checking RxJava by PMD 6.5.0 mentioned above. Although these five warnings are the same type, they pointed out different parts. Thus, they may have different lifetimes. We should not decide the lifetime of the warning by only one particular case.

(ii) *Warnings which alive a the end of the observation*: There are warnings which are alive in the latest version of the software, e.g., w_1 shown in Fig. 1. Since the death events of such warnings have not occurred until now, we cannot compute their actual lifetime.

To overcome the above two issues (i) and (ii), we propose an application of the survival analysis method [15] to the warning lifetime estimation. The survival analysis is a statistical analysis method for analyzing the expected duration of time until an occurrence of the target event. On the above issue (i), we regard each of warnings which are the same type of warning as a "sample" from the same population corresponding to the warning type. For example, we consider the above five "CommentSize" warnings to be five samples selected from the population of "CommentSize" warnings. Then, we statistically estimate the mean lifetime of the warnings of the type by the survival analysis method. We can overcome the remaining issue (ii) by using the survival analysis method as well. In the survival analysis, samples which have not experienced the event of interest (i.e., the death) until the end of the observation are referred to as "censored" samples, and there is a proper method to estimate the mean lifetime even though if we have censored samples. We explain a detailed way of estimation in the following Sect. 3.3.

3.3 Survival Analysis to Code Analyzer Warnings

The survival analysis is one of statistical analysis methods which are often used in the medical field. This method deals with the survival time of patients after they got a particular treatment: it expresses the probability that a patient survives at time t, as a survival function $S(t)$.

[6]Its file path is src/main/java/io/reactivex/internal/operators/flowable/FlowableRepeatWhen.java.

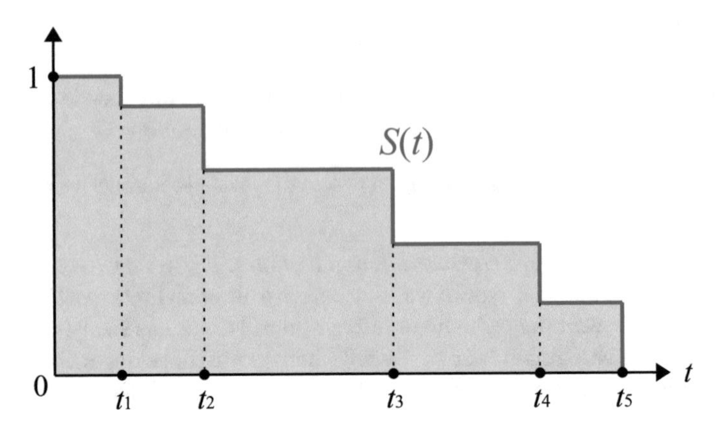

Fig. 2 An example of (discrete) survival function

At first, $t = 0$ and $S(t) = 1$. Then, $S(t)$ decreases over time by the occurrence of patient death. The area under $S(t)$ (see the gray region in Fig. 2) corresponds to the expected survival time (the mean lifetime), μ:

$$\mu = -\int t \, dS(t) = \int S(t) \, dt \; .$$

When we express the occurrence times of the death events by t_i, $S(t)$ is computed by the discrete hazard function $\lambda(t_i)$ (for $i = 1, 2, \ldots$); The hazard $\lambda(t_i)$ is the probability that the patient was alive at t_{i-1}, and dead at t_i, where $t_0 = 0$ and $\lambda(t_0) = 0$. For example, by using the hazard function $\lambda(t_i)$, we can express $S(t_2) = \{\, 1 - \lambda(t_1)\} \{1 - \lambda(t_2) \,\}$ because $S(t_2)$ is the probability that

(a) the patient survives at t_1, and
(b) the patient survives at t_2,

where the probabilities of (a) and (b) are $1 - \lambda(t_1)$ and $1 - \lambda(t_2)$, respectively. Hence, $S(t)$ is obtained as:

$$S(t_i) = \prod_{k=1}^{i} \{\, 1 - \lambda(t_k) \,\} \; .$$

By estimating $\lambda(t_i)$ from our data (samples), we can also estimate the survival function $S(t)$ and the mean lifetime μ.

In this paper, we apply this model to an analysis of code checker warnings: a warning corresponds to a patient, and warning death does to patient death, respectively.

In the estimation of hazard, we may face some censored samples, i.e., warnings which survive in the latest version of the source file (e.g., w_1 in Fig. 1), and we cannot get precise data of when they encounter the death events. There is a well-designed statistical method, Kaplan-Meier (KM) method [16], to estimate the hazard from data including censored samples. We can easily perform the estimation by the KM method with using `survfit` function included in `survival` package of R; R is an open source software for statistical computing.

In this study, we estimate the survival function for each type of warning. For the sake of convenience, we denote the survival function for the j-th type of warnings by $S_j(t)$, and do the expected lifetime by μ_j, respectively (for $j = 1, 2, \cdots$). We estimate $S_j(t)$ by using the KM method, and compute μ_j: the brevity of warning means its importance.

4 Case Study

In this section, we report our large-scale case study of code checker warnings through the survival analysis.

4.1 Aim

We collect warning data from many OSS projects, and perform a survival analysis of the collected data to understand the trends of code checker warnings including:

(i) how long warnings are alive in the real,
(ii) how many warnings are important and how many ones are not, and
(iii) how many disregarded warnings we can reduce by filtering based on our results.

4.2 Data Source

We collect warning data from 100 OSS Java projects which ranked in the top 100 in terms of "stars" at GitHub.[7] While we omit the list of projects due to space limitation, all data are available at http://se.cite.ehime-u.ac.jp/data/SCI2019/. The main reasons why we selected these projects are as follows.

- To collect data efficiently: We need all versions of all source files to trace the changes of all warnings which appear or had appeared in them. Thus, we have to access to the repository huge times. These projects use Git as their code repositories, and Git allows us to make a local copy of the repository, so we can get any version of any source file from our local copy of the repository without any network communication overhead.
- To get more informative results: We selected many projects as popular as possible to enhance the generality of our data analysis results. The more popular the project

[7]The ranking was as of July 17, 2018.

is, the more people are interested in it and the warnings appearing there would be better samples which reflect the trends in practice. Hence, we selected projects which have high scores in terms of the "stars" measure.

We used PMD 6.5.0 as the code checker. Since PMD is a popular OSS and anyone can use it freely, we believe that practitioners can easily utilize our results for their quality management. On the other hand, we have a restriction that our analysis is limited to Java software because of the PMD support language; further analyses of programs written in other languages are our future work.

4.3 Data Collection and Analysis Procedure

We conducted our data collection and analysis as follows.

(1) *Making local copies of repositories and lists of Java source files*: For all projects, we made local copies of their repositories, and built lists of latest Java source files.[8]
(2) *Tracing the change history of Java source files*: For each source file in the lists built at step (1), we traced back its change history (including a rename of the file path) to its initial version.
(3) *Collecting warnings*: We scanned all versions of all source files by using PMD, and collected data of warnings appeared in the source files. To avoid any bias on the rule set selection, we used all Java-related rule sets.[9] Since there can be two or more warning instances of the same kind within a source file, we traced their changes over commits separately: we liked each warning between two versions by the dynamic programming matching algorithm [17, 18].
(4) *Examining the birth events and the death events*: For all collected warnings, we examined the dates of birth and death and computed their lifetimes. For a censored sample, we substituted the death date by the date of the repository cloning (making a local copy).[10]
(5) *Performing the survival analysis*: For each kind of warnings, we performed a survival analysis and estimated its survival time (lifetime) as that kind of warning's importance (priority).

[8]We excluded a source file if its name ends with "Test.java" or its path includes "test," "demo," "sample," "example," or "template."

[9]They are the same as the example in Sect. 3.1.

[10]These are also required to estimate the hazard in the KM method.

4.4 Results

We collected 54,162 source files; since most of them have two or more different versions, we examined 1,130,009 different files in total. Through the PMD scanning and the warning-change tracing, we found 6,927,432 warnings (259 kinds of warnings). 3,113,649 out of 6,927,432 warnings (about 44.9%) were the censored samples.

Tables 2 and 3 show the results of survival analysis. From Table 2, we can see most types of warnings have long lifetimes; The median is about 1,446 days, i.e., about 4 years. Even the 25% percentile is about 760 days, i.e., about 2 years. Hence, most of the warnings tend to be left (not cleared) by programmers. While many types of warnings continue to appear in source files, a few types of them shown in the top

Table 2 Five-number summary of the expected lifetimes (in days)

Min	25%	50%	75%	Max
0.491	759.609	1445.696	2031.486	5442.567

Table 3 Warnings with the 10 shortest lifetime and 10 longest lifetime

Rank	Warning	Expected lifetime (in days)
1	AvoidLosingExceptionInformation	0.491
2	JUnitSpelling	5.522
3	StringBufferInstantiationWithChar	11.450
4	FinalizeOverloaded	12.882
5	CheckResultSet	13.634
6	EmptyInitializer	14.520
7	UseArraysAsList	19.268
8	FinalizeShouldBeProtected	20.593
9	UseEqualsToCompareStrings	21.552
10	EmptyTryBlock	64.111
⋮	⋮	⋮
250	AvoidSynchronizedAtMethodLevel	3140.768
251	GenericsNaming	3209.435
252	AtLeastOneConstructor	3309.299
253	CommentSize	3414.143
254	ClassNamingConventions	3440.574
255	AvoidUsingShortType	3577.736
256	TooManyMethods	3600.284
257	SingleMethodSingleton	3623.351
258	JUnit4TestShouldUseBeforeAnnotation	4424.262
259	JUnit4TestShouldUseAfterAnnotation	5442.567

of Table 3 have short lifetimes: only 9 warnings have lifetimes shorter than 1 month. Thus, these kinds of warnings are highly likely to be cleared sooner.

4.5 Discussion

We proposed to automatically evaluate the importance of warning using the survival analysis method. As a result of large-scale data analysis, most types of warnings have long lifetimes, i.e., they continue to appear for long periods. Since many programmers tend to leave them and not to fix, such warnings may be worthless in the code review. On the other hand, a few types of warnings have short lifetimes. For example, the mean lifetime of "AvoidLosingExceptionInformation" is about half a day, so it seems to be an important warning. Indeed, this warning is considered to be an error-prone pattern in PMD's rule sets. The Java-related rule sets in PMD consists of eight subsets: "Best Practices," "Code Style," "Design," "Documentation," "Error Prone," "Multithreading" "Performance," and "Security." In these subsets, "Error Prone" provides rules to detect constructs that are either broken, extremely confusing or prone to runtime errors. In our results, 8 out of the top 10 warnings shown in Table 3 are from "Error Prone" subset. On the other hand, in the worst 10 warnings, only one warning ("SingleMethodSingleton") is from that subset. While we used all Java-related rule sets, the survival analysis seems to reasonably evaluate such warnings which are of practical importance.

Notice that the above discussion does not mean we should use only "Error Prone" subset. Even if we use only the subset, there are 97 types of warnings. However, they do not always have short lifetimes as shown in Tables 2 and 3, i.e., they are not always considered to be important by programmers in the real world. Hence, just selecting only such a promising predefined rule set cannot be the best solution to the warning prioritization problem. Our results are derived from the trends of warnings in practice, so they reflect the real programmers' activities through the survival analysis. It can be a well-tuned solution, and moreover, code checker's users do not need to provide their feedback for the tuning.

Finally, we also check how many warnings we can reduce if we filter out by using our results. A warning having a long lifetime (e.g., over 1 year) would be worthless for many programmers in their code review activity. Figure 3 shows the cumulative appearance count of warnings over their expected lifetime (in the initial version of all source files). If we focus only on the warnings whose expected lifetime are shorter than 1 year, only 30 types of warnings remain, which are 11.6% of the original ones (reduced to 30 from 259 types); even if we do it with the ones whose lifetime are shorter than 2 years, the number of warning types is reduced to 22.4% (reduced to 58 from 259 types). When we focus on the total appearance count (not unique count) of warnings, the total number is reduced to only 0.2% of the original ones (reduced to 4,710 from 2,372,648) if we filter out the warnings whose lifetimes are longer than 1 year; that number is reduced to 1.3% (reduced to 30,427 from 2,372,648) even if we filter out the warnings whose lifetimes are longer than 2 years.

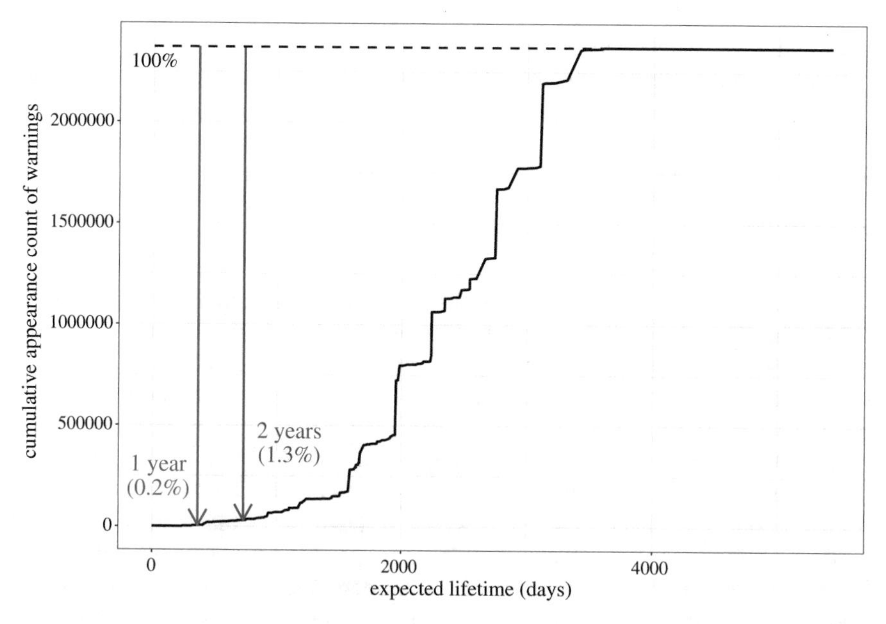

Fig. 3 The cumulative appearance count of warnings in the initial version of source files versus their expected lifetime

4.6 Threats to Validity

If a project enforced the developers to use a code checker or to obey to a coding rule, the trend of warnings in the project might differ from the other projects. For example, if a project required all source code and patches to pass a code checking by a specific checker before committing code, some kind of warnings would be cleared beforehand and do not appear in our case study. Since we did not examine all of the projects in great detail, the use of a code checker and the presence of a coding rule in a project are our internal threats to validity. Although we collected data from many different projects to observe a general trend, it is better to categorize projects from the above points of view.

Since our case study collected data from Java projects, the restriction of the programming language is an external threat to validity. We might observe a different trend of warning lifetimes from other-language projects. Since the programming practices are almost the same among the modern programming languages, we believe this threat would not be severe. To discuss the impact of the difference of the programming language, we need to perform a further study using other-language projects with a code checker other than PMD.

To trace changes of warnings, we performed the dynamic programming (DP) matching for the warning sequences between consecutive versions (commits). While the DP matching is a practical algorithm to compare two sequences, there is no guarantee that we completely traced all warnings without missing. For example, if a warning was cleared at a commit and another warning of the same kind was born near there, then we might misjudge the first warning continued to appear. We want to seek a better way in the future.

5 Conclusion

We proposed a survival analysis-based method for automatically evaluating the importance of code checker warnings. The proposed method utilizes the practical trends of warnings through code changes and does not require tool users to have any customization effort. Through a case study of 100 OSS projects, we showed that 259 types of warnings appeared and most of them tend to have remained unchanged for long periods; only 30 types of the warnings (11.6%) had lifetimes shorter than one year (data are available at http://se.cite.ehime-u.ac.jp/data/SCI2019/). Since shorter-lived warnings are related to the parts which programmers fixed sooner, we should notice such warnings to programmers with higher priority. On the other hand, longer-lived warnings would be worthless for programmers, and then we may assign lower priority to such warnings; if we filter out long-lived warnings whose lifetimes are longer than one year, we can reduce the total appearance count of warnings to only 0.2% of the original count. The proposed method would be useful in tuning the warning priority for more efficient utilization of code checkers.

Our future work includes: (1) a more detailed analysis of the relationship of warnings with the fault-introducing commits and an analysis of how short-life warnings affect the code changes; (2) data collections using other tools such as CODEBEAT and SonarQube, and comparisons of the results; (3) further analyses supporting many programming languages other than Java as well.

Acknowledgements This work was supported by JSPS KAKENHI #16K00099 and #18K11246. The authors would like to thank the anonymous reviewers for their helpful comments.

References

1. Jones, C.: Applied Software Measurement: Global Analysis of Productivity and Quality, 3rd edn. McGraw-Hill, New York (2008)
2. Li, Y., Li, D., Huang, F., Lee, S.Y., Ai, J.: An exploratory analysis on software developers' bug-introducing tendency over time. In: Proceedings of the 2016 International Conference on Software Analysis, Testing and Evolution, Yunnan, China, pp. 12–17, 3–4 Nov 2016. https://doi.org/10.1109/SATE.2016.9
3. Parnas, D.L.: Software aging. In: Proceedings of the 16th International Conference on Software Engineering, Sorrento, Italy, pp. 279–287, 16–21 May 1994
4. Cunningham, W.: The WyCash portfolio management system. ACM SIGPLAN OOPS Messenger **4**(2), 29–30 (1993). https://doi.org/10.1145/157710.157715
5. Brown, N., Cai, Y., Guo, Y., Kazman, R., Kim, M., Kruchten, P., Lim, E., MacCormack, A., Nord, R., Ozkaya, I., Sangwan, R., Seaman, C., Sullivan, K., Zazworka, N.: Managing technical

debt in software-reliant systems. In: Proceedings of the FSE/SDP Workshop on Future of Software Engineering Research, Santa Fe, New Mexico, pp. 47–52, 7–8 Nov 2010. https://doi.org/10.1145/1882362.1882373

6. Tufano, M., Palomba, F., Bavota, G., Oliveto, R., Penta, M.D., Lucia, A.D., Poshyvanyk, D.: When and why your code starts to smell bad (and whether the smells go away). IEEE Trans. Softw. Eng. **43**(11), 1063–1088 (2017). https://doi.org/10.1109/TSE.2017.2653105
7. Muske, T.B., Baid, A., Sanas, T.: Review efforts reduction by partitioning of static analysis warnings. In: Proceedings of the 2013 IEEE 13th International Working Conference on Source Code Analysis and Manipulation, Eindhoven, Netherlands, pp. 106–115, 22–23 Sept 2013
8. Johnson, B., Song, Y., Murphy-Hill, E., Bowdidge, R.: Why don't software developers use static analysis tools to find bugs?. In: Proceedings of the 2013 International Conference on Software Engineering, San Francisco, CA, pp. 672–681, 18–26 May 2013
9. Aggarwal, A., Jalote, P.: Integrating static and dynamic analysis for detecting vulnerabilities. In: Proceedings of the 30th Annual International Computer Software and Applications Conference, Chicago, Illinois, pp. 343–350, 17–21 Sept 2006. https://doi.org/10.1109/COMPSAC.2006.55
10. Hanam, Q., Tan, L., Holmes, R., Lam, P.: Finding patterns in static analysis alerts: improving actionable alert ranking. In: Proceedings of the 11th Working Conference on Mining Software Repositories, Hyderabad, India, pp. 152–161, 31 May–1 June 2014. https://doi.org/10.1145/2597073.2597100
11. Wang, S., Chollak, D., Movshovitz-Attias, D., Tan, L.: Bugram: bug detection with N-gram language model. In: Proceedings of the 31st IEEE/ACM International Conference on Automated Software Engineering, Singapore, pp. 708–719, 3–7 Sept 2016. https://doi.org/10.1145/2970276.2970341
12. Burhandenny, A.E., Aman, H., Kawahara, M.: Examination of coding violations focusing on their change patterns over releases. In: Proceedings of the 23rd Asia-Pacific Software Engineering Conference, Hamilton, New Zealand, pp. 121–128, 6–9 Dec 2016. https://doi.org/10.1109/APSEC.2016.027
13. Shen, H., Fang, J., Zhao, J.: EFindBugs: effective error ranking for FindBugs. In: Proceedings of the 4th IEEE International Conference on Software Testing, Verification and Validation, Berlin, pp. 299–308, 21–25 Mar 2011. https://doi.org/10.1109/ICST.2011.51
14. Sadowski, C., Gogh, J.V., Jaspan, C., Söderberg, E., Winter, C.: Tricorder: building a program analysis ecosystem. In: Proceedings of the 37th International Conference on Software Engineering, Florence, Italy, pp. 598–608, 16–24 May 2015
15. Rupert, J., Miller, G.: Survival Analysis. Wiley, Hoboken, New Jersey (2011)
16. Kaplan, E.L., Meier, P.: Nonparametric estimation from incomplete observations. J. Am. Stat. Assoc. **53**(282), 457–481 (1958). https://doi.org/10.1080/01621459.1958.10501452
17. Cormen, T.H., Leiserson, C.E., Rivest, R.L., Stein, C.: Introduction to Algorithms, 3rd edn. MIT Press, Cambridge, MA (2009)
18. Sakoe, H., Chiba, S.: Dynamic programming algorithm optimization for spoken word recognition. IEEE Trans. Acoust. Speech Signal Process. **26**(1), 43–49 (1978). https://doi.org/10.1109/TASSP.1978.1163055

Elevator Monitoring System to Guide User's Behavior by Visualizing the State of Crowdedness

Haruhisa Hasegawa and Shiori Aida

Abstract Internet of things (IoT) is expected to make our social life more convenient. In the IoT, various lifestyle tools measure the surrounding situation using sensors that send the data to a cloud. The cloud analyzes a set of the data, visualizes it and/or sends feedback to the "things." However, there are many old facilities around us that were established in the past and do not have a sensing mechanism or the ability to send data, and so they become isolated from the concept of IoT. In this paper, we show that using an elevator for an example, even old equipment can be made efficient using IoT. We propose an IoT system that improves the fairness and efficiency by visualizing the crowdedness of an elevator, which has only one cage. When a certain floor gets crowded, unfairness arises in the users on the other floors as they are not able to take the elevator. Our proposed system improves the fairness and efficiency by guiding the user's behavior. The edge device collects information such as the existence and the destination floor of users using beacon technologies. The cloud predicts the crowdedness based on the data transferred by MQTT. Additionally, this system does not require the elevator to have connectivity to the internet. Further, the simulation results of the effect of the proposed system are shown.

Keywords IoT · Cyber physical system · Elevator control system · Beacon · Simulation · MQTT

H. Hasegawa (✉) · S. Aida
Department of Mathematical and Physical Sciences, Japan Women's University, 2-8-1 Mejirodai, Bunkyo-Ku, Tokyo 112-8681, Japan
e-mail: hasegawah@fc.jwu.ac.jp

S. Aida
e-mail: m1216001as@ug.jwu.ac.jp

R. Lee (ed.), *Big Data, Cloud Computing, and Data Science Engineering*, Studies in Computational Intelligence 844,
https://doi.org/10.1007/978-3-030-24405-7_6

1 Introduction

Information and communication technology (ICT) is being widely applied throughout various public infrastructures that are connected to monitoring systems. The operator monitors the status of the system and arranges for repairs to be conducted when a failure or abnormality is detected. For example, advanced elevators are monitored by an integrated operation center so that safety and security are guaranteed and the efficiency of operation is improved. However, the system does not provide users with any convenience as the information is not disclosed.

The latest elevator control system (ECS) [1] in a tower building is controlled by a sophisticated algorithm that optimizes the movement of multiple cages. Other public systems, which have been used for a long time, are controlled by old and primitive algorithms. It is common that old public systems do not have internet or intranet access and so it is difficult to link them with external applications. Therefore, the elevator can only be controlled based on the information that it senses.

The internet of things (IoT) is one of the most important recent IT trends. Recently, IoT technologies are finding applications in various file and are expected to exchange more data than before. An IoT edge device collects data from the surrounding environment using sensors, and then sends it to a cloud via the internet. The cloud integrally analyzes and visualizes the data sent from multiple edge systems. Recently, an IoT system is often utilized for collecting big data. Contrastingly, a Cyber Physical System (CPS) [2] aims to mutually realize the interaction between edge devices and the cloud. An ECS has the potential to be more sophisticated based on the concept of a CPS.

An elevator group control (EGC) [3] system has been applied to ECS to efficiently transport users. Recently, a call assignment control method has become mainstream. In this system, a "hall call" is registered as soon as the user presses the hall call button. Then, the elevator cage, the one that is judged by the EGC to be the most appropriate within the group, is assigned to the hall call. Nowadays, artificial intelligence technology is used for optimum cage assignment. Thus, in the case of an elevator system with plural cages, it is possible to apply sophisticated optimization. However, if an elevator system has only one cage as seen in relatively old buildings, there is a limit to the efficiency of the operation.

We propose a system that guides the movement of users according to the crowdedness on each floor and cage. It improves transportation efficiency by presenting information such as the current usage situation based on CPS. Herein, a technique that makes elevator's operation more fair and efficient is proposed. It is designed for a single elevator with only one cage based on IoT/CPS. It is usually very difficult to add a new function to ECS, such as the IoT function, because the system is constructed as a closed system. Our proposed system does not require the ECS to use a modified algorithm or to have new capabilities such as accessing an operation center and/or the internet. When our proposed system is attached to the elevator in operation, there are no problems even if the ECS is left unmodified. Herein, we introduce the system architecture and implementation regarding the proposed technique. Additionally, we validate the improvements in fairness and efficiency by using a simulation of its performance.

2 Problems of a Single Cage Elevator

There are some examples of an IoT system that notifies many unspecified users of the collected information to reduce the inconvenience of public infrastructure. For example, the visualization of the next arriving congested train [4]. The users are able to decide their behavior by considering the degree of congestion. This is expected to reduce congestion because users who are not in a hurry will choose a relatively less crowded train. A similar effect is expected for the operation of an elevator. Regarding an elevator with a single cage, usually waiting users are shown what floor the cage is at and in which direction it is moving. However, the users are not able to figure out the number of people waiting on the other floors or riding in the elevator. The users must decide whether to wait for the elevator or forgo it from the limited information they have. Because the user can estimate the level of congestion only when the elevator arrives, there is a large delay the decision making. Even then, the user is not able to get the information about other floors. Therefore, we propose that the congestion status on each floor be disclosed to users to enable them in choosing an appropriate action.

For a single-cage elevator system, it is crucially important to solve the problem that users on other floors cannot get on the elevator when a previous particular floor is crowded. Especially, for old single cage elevators that are operated using a primitive algorithm [5]. For this problem, we visualize how the users should behave to efficiently get on the elevator based on an IoT/CPS architecture. The IoT system collects data of the number of users waiting on each floor and their destination floor and sends this information to the IoT platform using the MQTT protocol. The IoT platform notifies the users that there is the possibility of getting on the elevator earlier than at the floor he/she is now waiting at, i.e., the system recommends users to move to neighboring floors via the stairs.

One proposed technique to improve the efficiency of an elevator with multiple cages is to register the floors that each user wants to go while they are waiting [6]. In this case, it is efficient to guide only the users that have the same destination floor to use the same elevator thus reducing the number of times the elevator stops. When the elevator is almost full, the number of unnecessary stops decreases. If the number of stops at which only a few people can board the elevator decreases, the elevator operation will be more efficient. However, a conventional elevator with a single cage, which is the focus of this paper, is simply operated according to a "hall call" from each floor. One problem with single cage elevators is the unfairness between the users at different floors.

For example, when the elevator is moving upward, it must move to the highest floor, which has been demanded by the users in the elevator. Only after it arrives at the designated floor can the elevator descend. If there are more users than its capacity waiting at a floor that the elevator passes through, they are not able to board. Even if the upper floor is not crowded, the situation is the same. Therefore, we propose an IoT system to visualize the possibility that users are able to board the elevator at each floor. With reference to the information provided, if some users move to another floor via the stairs, they are able to get on earlier than waiting on the original floor.

3 Proposed System

The basic architecture of the proposed system is shown in Fig. 1. It is assumed that users of the elevator carry their own terminal such as a smartphone or tablet. Each terminal periodically transmits a signal based on personal area network (PAN) technologies, as a beacon. Each terminal sends its MAC address as an identifier via a 2.4 GHz band [7, 8], such as IEEE 802.15.4 [9] or IEEE 802.15.1 [10]. Additionally, each user also inputs their intended destination floor. There are sensor devices installed in the elevator hall of each floor and in the elevator.

The sensor device counts the number of terminals in each elevator hall by identifying the beacon's MAC address. It is well known that the received signal strength indicator (RSSI) of a beacon is inversely proportional to the square of the distance. Therefore, the sensor device is able to judge if the terminal is within the predetermined area or not. However, it is difficult to accurately estimate the distance because in general the reflection and interference of signal is large. Figure 2 shows the theoretical calculation result of the RSSI, which is affected by reflection from the ground. The frequency was set to 2.4 GHz, and the ground heights of the transmitting and receiving antennas were set to 0.6 m and 0.62 m, respectively. As seen from Fig. 2, the RSSI greatly differs even with a slight difference in distance. But Fig. 2 also shows that it is possible to estimate if the terminal is located within about 1 m by setting the threshold to ~40 dbm. Therefore, the number of passengers in the elevator can be estimated. Further, by setting the threshold at ~60 dbm, it is able to estimate if the terminal is within the elevator hall. The sensor device then notifies the IoT platform of the set of MAC addresses via WAN. Many narrowband mobile network services [11–13] dedicated for IoT have become available in recent years. Because only text information is transferred in the proposed system, these services are suitable and cost-effective. The sensor devices do not communicate with the IoT platform via the ECS because we have focused on relatively older equipment. Additionally, it is desirable that the WAN is a closed network because it transfers the user's identifier. On the IoT platform, it is possible to predict the congestion in the elevator from the information gathered, which can then be passed on to the users waiting on each floor. The IoT platform notifies the user's terminal via the internet. Because this informa-

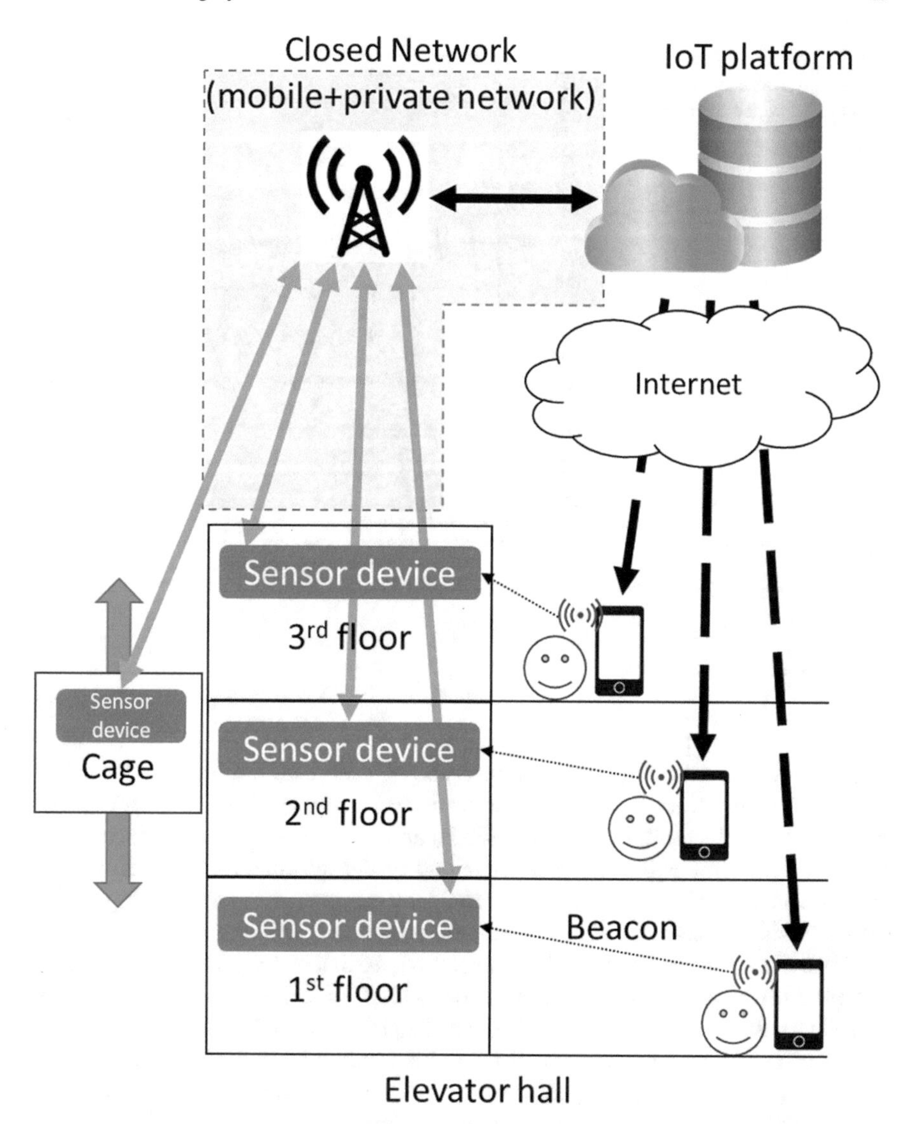

Fig. 1 Architecture of proposed system

tion does not relate to an individual user, it can be transferred by an open network. The users are then urged to move to another floor if it is difficult for them to board the coming elevator.

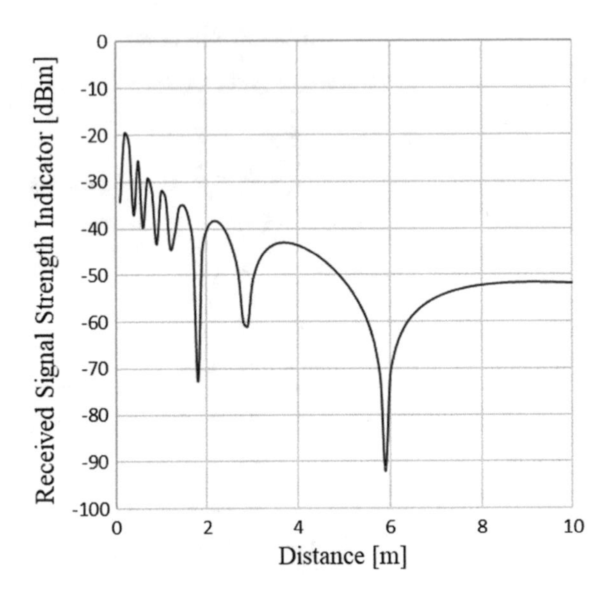

Fig. 2 Received signal strength indicator of beacon with the influence of earth reflection

4 Implementation

In the proposed system, it is necessary for the sensor device to discover the existence of arbitrary user terminals. As explained above, each user terminal broadcasts its own MAC address and destination floor via the beacon technology using the 2.4 GHz band, such as Bluetooth Low Energy (BLE) and ZigBee. The sensor device placed on each floor notifies the IoT platform/cloud of the information it has discovered from the user terminals. A sequence for realizing this process is shown in Fig. 3. The sensor device manages the list of MAC addresses on each floor and notifies the IoT platform of any newly discovered beacons and those that no longer exist. Because the IoT platform must manage the information, which is received asynchronously, it is suitable for the data to be transmitted via MQTT [14]. Further, because the transferred information relates to the location of the users, sending it via the internet is unsafe. The sensor device and the IoT platform are connected by a mobile telecommunication service without depending on the network equipment in the building. Because the information is only text data of the user's address and so on, it is possible to use narrowband mobile communication or low power wireless access (LPWA) services. In the IoT platform, the received data is registered on a relational data base (RDB). The application and WEB servers visualize and notify the users of the crowdedness in real time based on the obtained data. The beacon terminal does not have to be the same as the terminal that receives the notification from the IoT platform. The beacon device should be what the users always carry with themselves. If there are undetected users on each floor of the system, the estimation of the congestion situation becomes inaccurate. It is possible that the beacon could be installed into the user's employee ID card, admission/visitor passes, and so on. Conversely, the user terminal receiving the

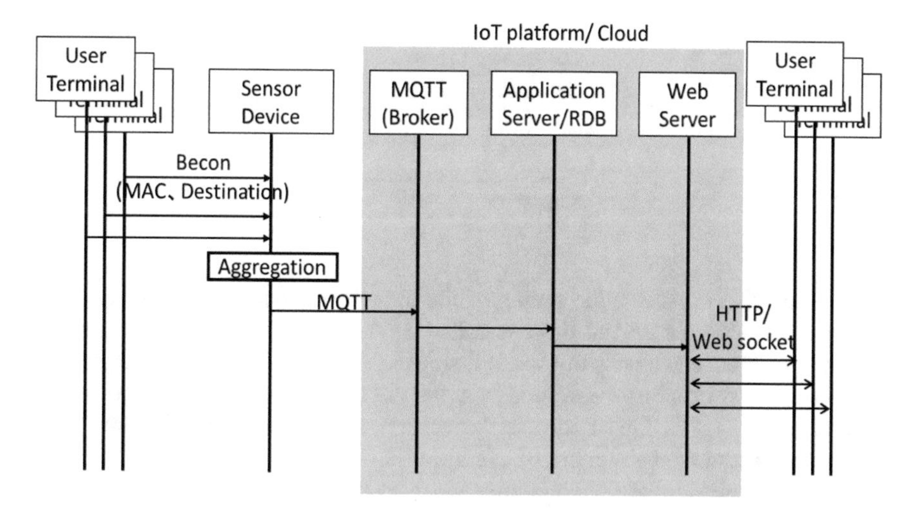

Fig. 3 Sequence diagram of proposed system

notification must be able to visualize the information it receives and should therefore be a smartphone or a tablet. However, even if all users do not have such a device, the effect will be reduced; however, the system can still operate.

5 Simulation

We have evaluated the efficiency and fairness of the proposed system by simulation of elevator's movement and user's behavior. A simulation was performed based on a multi-agent simulation method, which was suitable for modeling the behavior of the elevator and the users. In the simulation, all users behave based on the same agent model, but they act differently depending on the situation of the floor and the cage. Each user wants to move to another floor after random time. An elevator moving in the upward direction was simulated according to the conditions shown in Table 1.

The building was five stories high and the capacity of the elevator was set to ten. The number of users on each floor was determined to be the initial state of the simulation. We assumed that there are users only on the first and the second floor, who want to move to an upper floor according to a random but predetermined

Table 1 Simulation conditions

Capacity of cage	10 persons
Building	1–5F
Destination floor (1F user)	3–5F
Destination floor (2F user)	4–5F

Table 2 Step value of operations

Door opening operation	2 step
Door closing operation	2 step
Start moving	1 step
Move the floor	2 step × floor
Using stairs	14 step

probability. Table 2 shows the setting of the time for the operation of the elevator. In this simulation the unit of time, is called a "step." As there are many users on the first floor, the elevator is full when it stops at the second floor. This situation is unfair for users. Users on the second floor are not able to get on the elevator unless all of the users on the first floor user have been transported to their desired floor. The system acquires the number of users waiting on the first floor at the entrance. When it is predicted that the number of users on the floor is too many for them to be able to take the next elevator, the system notifies the users of which floor they can take the elevator. In this simulation, the system does not notify the users about the crowdedness of the elevator and lower floors because the other floors will be crowded if users on the first floor move to other floors simultaneously.

The number of users on the second floor was set to 30, 60, and 80. The percentage of users on the first floor of the total users was changed. In the initial state, all users are on only the first and second floors. When 100% of all users are on the first floor, there are no users on the second floor. At 50%, the numbers of users on the first and the second floor are the same. The time until the last user on each floor reaches their destination floor is shown in Figs. 4, 5, and 6. As the total number of users increases, the time also increases. When the system is not in operation, the time on the second floor is always longer than on the first floor and so it is unfair for the second floor user. Conversely, our proposed system reduces the time on the second floor when the first floor is more crowded. When the percentage of the users on the first floor is more than or equal to 80%, the time on the second floor becomes less than on the first floor. This means that the user on the second floor does not need to wait until all the users on the first floor reach their destination floor. The effect is more apparent as the total number of users increases. According to Fig. 6, the time on the second floor becomes shorter even when the percentage of the first floor users is about 20%. However, the time on the first floor does not increase.

Figures 7, 8, and 9 show the average time it takes for users of each floor to reach their destination floor. Even in this evaluation, when the system is not operated, the time on the second floor is always longer than that on the first floor. The proposed system also reduces the average values on the second floor. It is more apparent, when the percentage of users on the first floor becomes more than or equal to 50%, i.e., when the number of users on the first floor becomes more than or equal to the number of users on the second floor. The proposed system does not increase the average time that users on the first floor reach the destination floor by much. Comparing Figs. 4, 5,

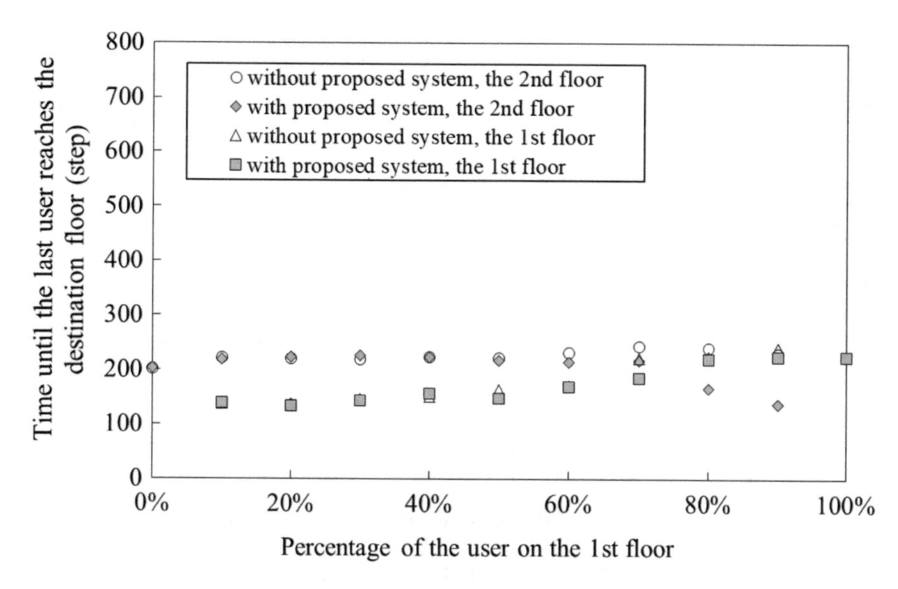

Fig. 4 Time until the last user reaches the destination floor (Number of the second floor user is 30)

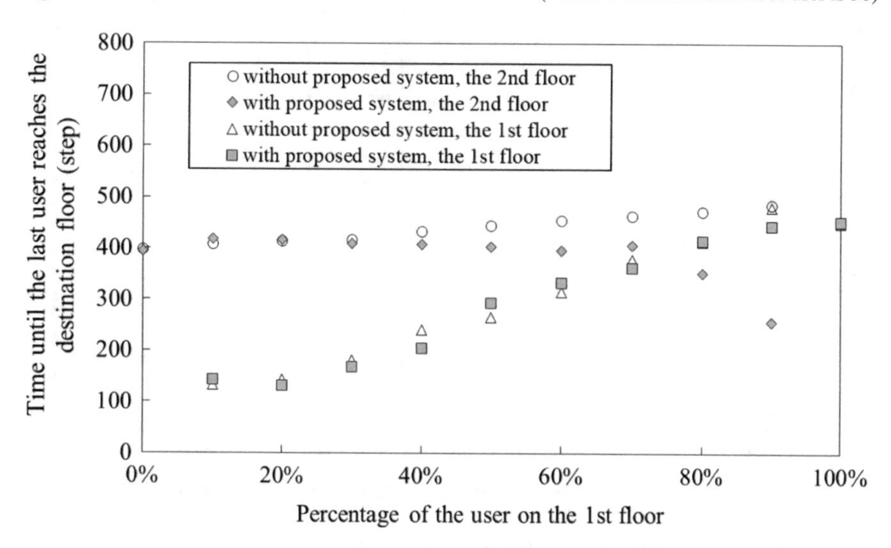

Fig. 5 Time until the last user reaches the destination floor (Number of the second floor user is 60)

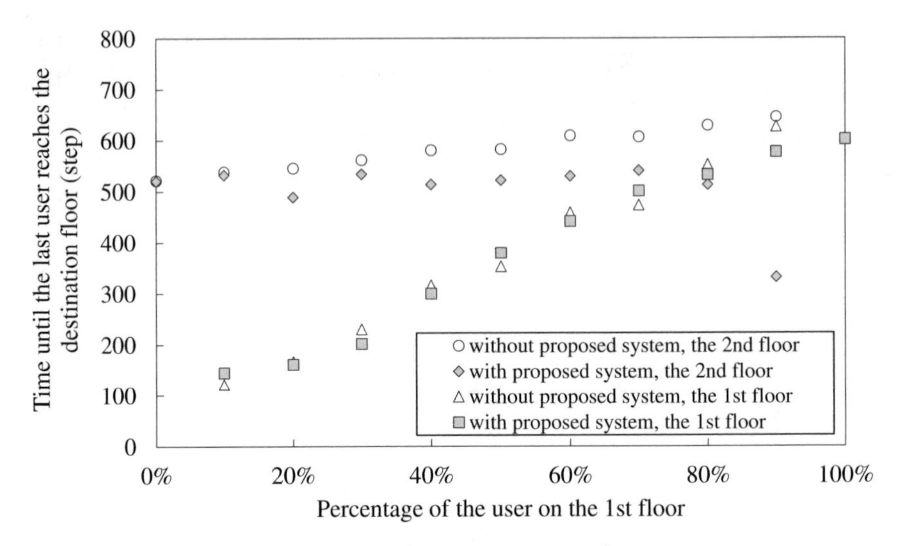

Fig. 6 Time until the last user reaches the destination floor (Number of the second floor user is 80)

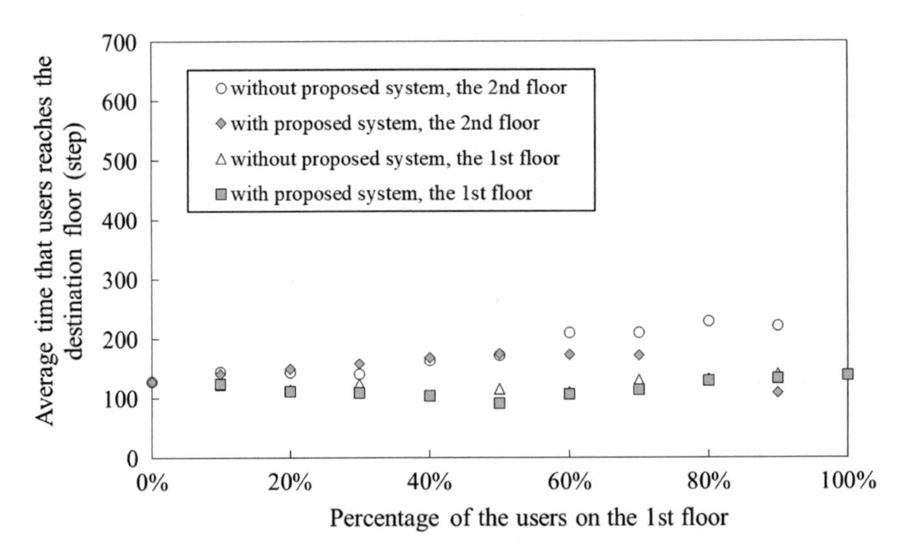

Fig. 7 Average time to reach the destination floor (Number of the second floor user is 30)

6, 7, 8 and 9, the longest and average time of users on the second floor are relatively similar when the system is not operated. This means that almost all users on the second floor must wait for the first floor users to reach their destination floor.

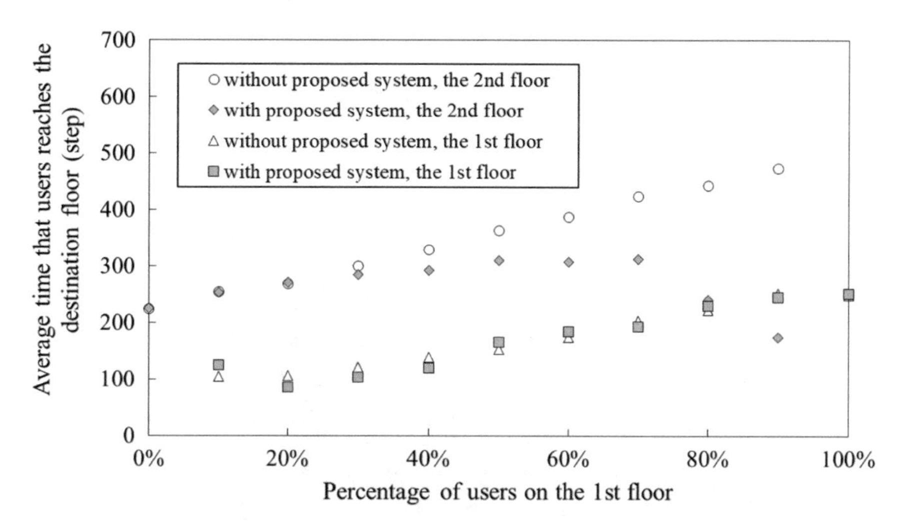

Fig. 8 Average time to reach the destination floor (Number of the second floor user is 60)

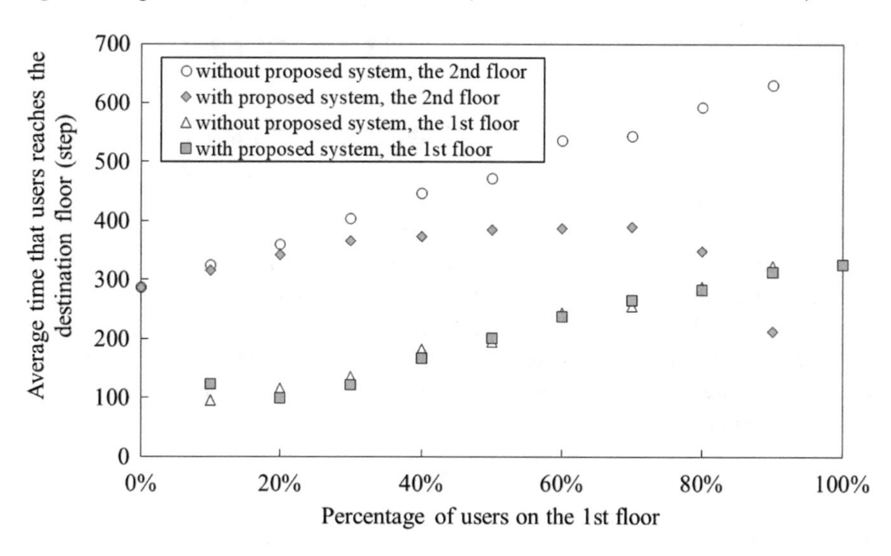

Fig. 9 Average time to reach the destination floor (Number of the second floor user is 80)

Figure 10 shows the histogram of the time taken to reach the destination floor when the proposed system is not operated. The number of users on the second floor was set to 60. The percentage of users on the first floor of the total users was set to 90%. The vertical axis denotes the number of users and the horizontal axis denotes the required time step. The hatched area shows the number of the second floor user. According to Fig. 10, it can be seen that the users on the second floor were not able to get on the elevator unless all the users on the first floor had been transported. The users on the second floor had to wait for the all the users on the first floor to

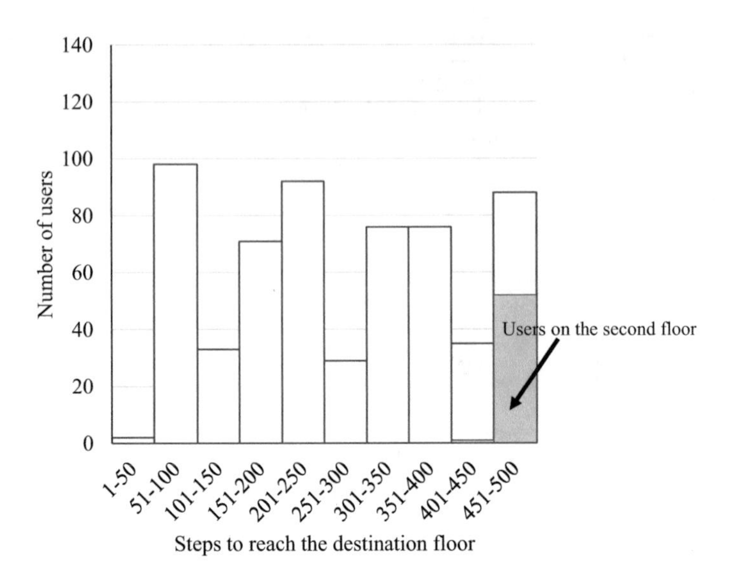

Fig. 10 Histogram of time to reach the destination floor without the proposed system operated

reach their destination floor. It is obvious that there is unfairness between a lower crowded floor and the upper floors. Figure 11 displays the histogram for when the proposed system is operated. According to Fig. 11, it can be seen that the users on the second floor are not required to wait until all the users on first floor user have been transported; therefore, the efficiency has improved. Some of the users on the second floor succeeded in taking the elevator at the beginning. It is clear that the fairness on the second floor than the first floor improves by applying the system. Additionally, it can be seen from Fig. 11 that the time until all the users arrive at their destination floor is also reduced. The efficiency of elevator use has been improved by guiding users to move to other floors.

6 Summary

In this paper, we proposed an information disclosure system based on IoT/CPS for elevators that have only one elevator. This system grasps the number of users waiting on each floor, predicts the crowdedness of the elevator, and guides the users on other floors appropriately. Therefore, the system makes the elevator's operation more fair and efficient. Here we presented both the architecture and how the system can be implemented. We adopted beacon technologies to know how many users are waiting on each floor. The data is transferred by MQTT via a mobile narrow-band service to the IoT platform. We also confirmed how the fairness and efficiency increases using this system. From the simulation results, the fairness and efficiency are seen

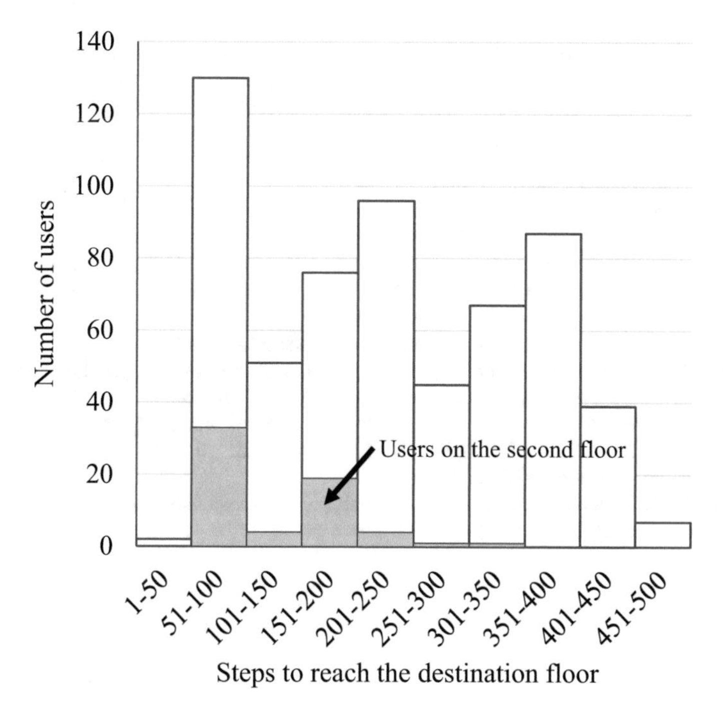

Fig. 11 Histogram of time to reach the destination floor with the proposed system operated

to have improved. If the proposed system is operated when the number of people on the crowded floor is larger than other floors, the time spent waiting on the second floor becomes shorter and unfairness is reduced. The fairness becomes better in the case where the proportion of users at the congested floor exceeds 50% of all users. Conversely, the waiting time for users at the crowded floor does not get any worse.

References

1. Fujino, A., Tobita, T., Segawa, K., Yoneda, K., Togawa, A.: An elevator group control system with floor-attribute control method and system optimization using genetic algorithms. IEEE Trans. Ind. Electron. **44**(4), 546–552 (1997)
2. Rajkumar, R.: Cyber-physical systems: the next computing revolution. In: Proceedings of the 47th Design Automation Conference, USA (2010, July)
3. Fernandez, J.R., Cortes P.: A survey of elevator group control systems for vertical transportation: a look at recent literature. IEEE Control. Syst. Mag. **35**(4) (2015, August)
4. Matsumoto, T., Nakagawa, T.: Train net onboard information service for individuals. JR EAST Tech. Rev. (24)-Autumn 2012, 15–20 (2012)
5. Strakosch, G.R., Caporale, R.S.: The Vertical Transportation Handbook (4th). Wiley (2010, November)

6. Beielstein, T., Ewald, C.-P., Markon, S.: Optimal elevator group control by evolution strategies. In: Proceeding of the 2003 International Conference on Genetic and Evolutionary Computation, vol. 2, pp. 1963–1974 (2003)
7. Benkic, K., Malajner, M., Planinsic, P., Cucej, Z.: Using RSSI value for distance estimation in wireless sensor networks based on ZigBee. In: 2008 15th International Conference on Systems, Signals and Image Processing (2008, August 22)
8. Botta, M., Simek, M.: Adaptive distance estimation based on RSSI in 802.15.4 network. Radioengineering, **22**(4), 1162–1168 (2013, December)
9. Institute of Electrical and Electronics Engineers: IEEE Std. 802.15.4-2003 Wireless Medium Access Control (MAC) and Physical Layer (PHY) Specifications for Low Rate Wireless Personal Area Networks (LR-WPANs) (2003, October)
10. Bluetooth Special Interest Group: Specification of the Bluetooth System
11. GPP TR 36.802: Narrowband Internet of Things (NB-IoT). Technical Report TR 36.802 V1.0.0, Technical Specification Group Radio Access Networks (2016, June)
12. LoRa Alliance: LoRaWAN What is it. Technical Marketing Work-group 1.0 (2015, November)
13. Raza, U., Kulkarni, P., Sooriyabandara, M.: Low power wide area networks: an overview. IEEE Commun. Surv. Tutor., 1–15 (2017)
14. IBM MQTT Protocol Specification. http://public.dhe.ibm.com/software/dw/webservices/ws-mqtt/mqtt-v3r1.html

Choice Behavior Analysis of Internet Access Services Using Supervised Learning Models

Ken Nishimatsu, Akiya Inoue, Miiru Saito and Motoi Iwashita

Abstract The purpose of this study is to understand the Internet-access service choice behavior considering the current market in Japan. The customer number of high-speed wireless services in Japan is growing rapidly in recent years. The choice behavior in the Internet-access service market is becoming complicated and diversified. We focus on two segments of Internet access users: fixed-line users and only-wireless users. Fixed-line users mean the customers use both fixed-line services and wireless services at home. Only-wireless users mean the customers use only-wireless services at home. We analyzed the differences between two user segments: fixed-line users and only-wireless users from various viewpoints on the basis of an original survey. We propose supervised learning models to create differential descriptions of these user segments from the viewpoints of decision-making factors. The characteristics of these user segments are shown by using the estimated models.

Keywords Demand analysis · Service choice behavior · Internet access services · Internet access behavior · Supervised learning model

K. Nishimatsu (✉)
NTT Network Technology Laboratories, NTT Corporation, Musashino, Japan
e-mail: ken.nishimatsu.hd@hco.ntt.co.jp

A. Inoue · M. Saito · M. Iwashita
Chiba Institute of Technology, Narashino, Japan
e-mail: akiya.inoue@it-chiba.ac.jp

M. Saito
e-mail: s1242040YF@s.chibakoudai.jp

M. Iwashita
e-mail: iwashita.motoi@it-chiba.ac.jp

R. Lee (ed.), *Big Data, Cloud Computing, and Data Science Engineering*,
Studies in Computational Intelligence 844,
https://doi.org/10.1007/978-3-030-24405-7_7

1 Introduction

The Internet usage rate (personal) was 80.9% in Japan. The smartphones to access the Internet accounted for 59.7% of all Internet users, which exceeded 52.5% for computers [1]. The household ICT device ownership rate of 2017 was 94.8% for mobile terminals and 72.5% for computers. The rate for smartphones, which are included in mobile-terminals category, increased to 75.1% exceeding the household ownership rates of computers [1].

Broadband services are available for almost all homes in Japan. Due to the improvement of wireless technologies such as LTE and WiMax services, various changes occur around the Internet market. The customer number of high-speed broadband services as of September, 2018 is indicated in Table 1 [2]. The high-speed fixed-line services include Optical-Fiber and high-speed CATV services. The minimum speed of high-speed CATV services is more than 30 Mbps. The high-speed wireless services include LTE and BWA (*WiMAX*, etc.) services. The customer number of high-speed wireless services in Japan is growing rapidly in recent years. It was common sense that Internet users use fixed-line services such as Optical-Fiber line at home, and wireless-line services outdoors. Internet users using the wireless services both at home and outdoors are increasing recently. And besides, new products, home routers were introduced in Japan. Home routers have multiple LAN ports and wireless lines of which the number of connections with terminals is more than 30, and are used with AC power supply. Therefore, some of users cancel the fixed-line broadband services used at home. On the other hand, there are heavy Internet users that need high and stable data transmission speed, and would like to use fixed-charge services with no usage limitation. New Optical-Fiber services of which the maximum speed is 10 Gbps was introduced in Japan. And the bundle service of fixed-line Internet access service and smartphones prevents the decrease of the fixed-line service users in Japan recently. Under such circumstance, it is a very important issue to understand the choice behavior of Internet-services for the related providers including mobile carriers, Internet service providers, Internet access line providers, contents providers and so on.

Table 1 Customer number of high-speed broadband services

Type of services		Number of customers (million)
High-speed fixed-line services		34.63
	Optical fiber	30.71
	CATV (down speed: more than 30 Mbps)	3.92
High-speed wireless services		189.53
	LTE	127.72
	BWA (WiMAX, etc.)	61.81

We proposed a Framework for Scenario Simulation to analyze market structure and estimate service demand [3–9]. The choice modeling is the most important component of the Framework. We constructed various types of choice models to analyze customer preference and understand customer choice behavior in the Internet service market and the mobile phone market in Japan [3–15]. The purpose of this study is to understand the Internet-access service choice behavior considering the current market in Japan. The choice behavior in the Internet-access service market is becoming complicated and diversified under above-mentioned circumstance. It is very difficult to construct Internet-access service choice models based on only the charge and the performance. And it is very difficult to define the choice set in Internet-access service choice behavior. There are various types of choice sets in the following. A user contracts with a mobile carrier to use the bundle service of fixed-line Internet access service and smartphones. Another user contracts with one of Internet-access line providers and one of Internet service providers. The choice sets and their decision-making factors vary from uses to users.

We focus on two segments of Internet access users: *fixed-line users* and *only-wireless users*. *Fixed-line users* mean the customers use both fixed-line services and wireless services at home. *Only-wireless users* mean the customers use only-wireless services at home. We analyzed the differences between two user segments: *fixed-line users* and *only-wireless users* from various viewpoints on the basis of an original survey conducted in January 2018. The book, *Data Science for Business* [16] describes a way to create differential descriptions of the clusters using supervised learning in the section, *Using Supervised Learning to Generate Cluster Descriptions*. The objective of this paper is learning the differences of user segments divided by observable variables, for example, current Internet access services. Therefore, we propose supervised learning models to create differential descriptions of the user segments in this paper.

2 Summary of Survey

The original survey was conducted to analyze the preference for the customer choice behavior for internet access services in January 2018. Sample data were collected by using the Web interview system provided by NTT Com Online Marketing Solutions Co. The sampling was carried out on the basis of the following requirements.

- Business customers are excluded.
- The customers are limited to decision-makers to choose services and their carriers in their household.
- The mobile customers are not only three mobile-carriers: *NTT docomo*, *au by KDDI*, and *SoftBank*, but also *MVNO* customers including SIM-free phone users.
- The number of individuals for only-wireless users is collected as much as possible.

There is no requirement for samples in relation to the other demographic factors such as age, gender, area, income and occupation. The number of individuals in

Table 2 Sample sizes by gender category

Gender	Number of samples	Share (%)
Male	764	64.3
Female	424	35.7
Total	1188	100

Table 3 Sample sizes by age category

Age	Number of samples	Share (%)
10s	1	0.1
20s	38	3.2
30s	210	17.7
40s	341	28.7
50s	329	27.7
Over 60s	269	22.6
Total	1188	100

Table 4 Sample sizes by user segment

User segment	Number of samples	Share (%)
Fixed-line users	749	63.0
Only-wireless users	439	37.0
Total	1188	100

the sample by gender category and age category are shown in Table 2 and Table 3, respectively.

To analyze the user preference for internet access services at home, we define two user segments: *fixed-line users* and *only-wireless users*. *Fixed-line users* indicate the users who use fixed broadband services such as Optical-Fiber, CATV and ADSL. *Fixed-line users* include users who use both fixed-line services and wireless services. *Only-wireless users* indicate the users who only use wireless services at home. Table 4 shows the sample size of each segment.

3 Characteristics of Internet Users

Figure 1 shows the mobile terminals that *only-wireless users* use mainly to access the Internet at home. About 60% of *only-wireless users* access the Internet by using mobile phone terminals (such as smartphones) or tablet. Figure 2 and Fig. 3 show histogram of the Internet-user number for fixed-line users and only-wireless users, respectively. It indicates that the number of the Internet users in the same household varies according to user categories and the ratio of single person households in only-wireless users is larger than that in fixed line users.

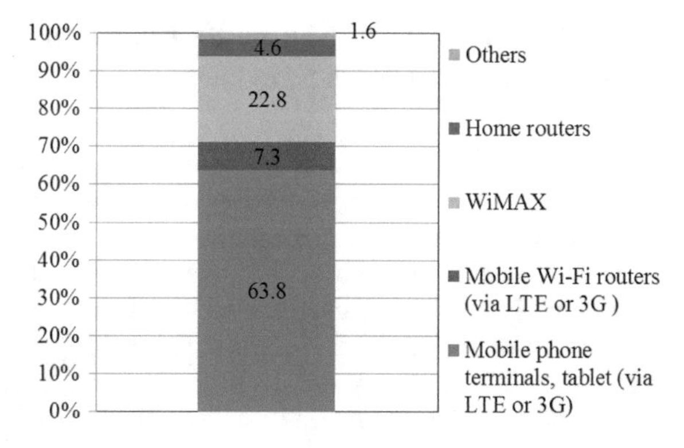

Fig. 1 Main mobile terminal of only-wireless users at home

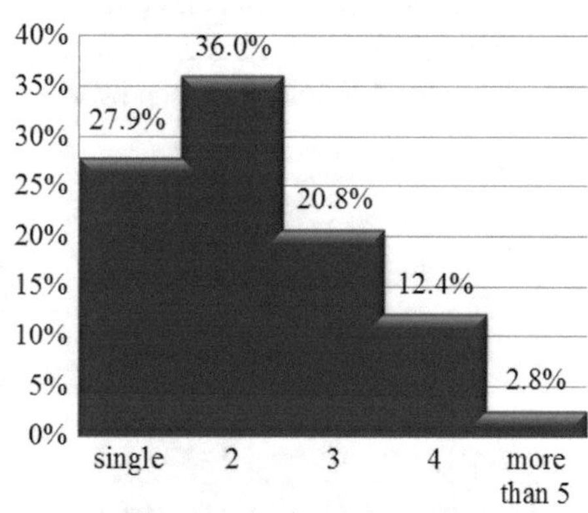

Fig. 2 Number of Internet users in same household, belonging to fixed-line users

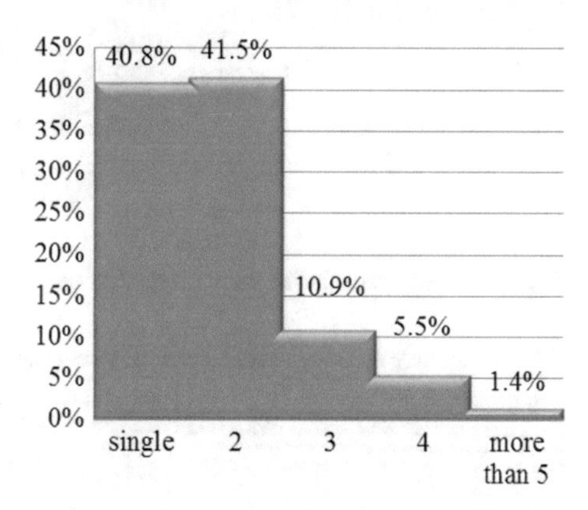

Fig. 3 Number of Internet users in same household, belonging to only-wireless users

4 Definition of User Segments Based on Use Intention for Internet Access Services

We analyze the Internet access service choice behavior of two customer segments shown in Table 4 in this paper. We asked these users whether or not you will plan to contract with other services in the future. Furthermore, we classify each user segment into two segments: stable users and unstable users shown in Fig. 4 on the basis of the results. The definition and the sample sizes by *fixed-line users* and *only-wireless users* are shown in Table 5 and Table 6, respectively. It is found that the rate of users who would like to or may use the current service in the future is more than 60%, though the rate of *fixed-line users* is larger than that of *only-wireless users*.

We focus on the following four user segments: stable *fixed-line users*, stable *only-wireless users*, unstable *fixed-line users* and unstable *only-wireless users* in order to understand the differences among these segments in this paper.

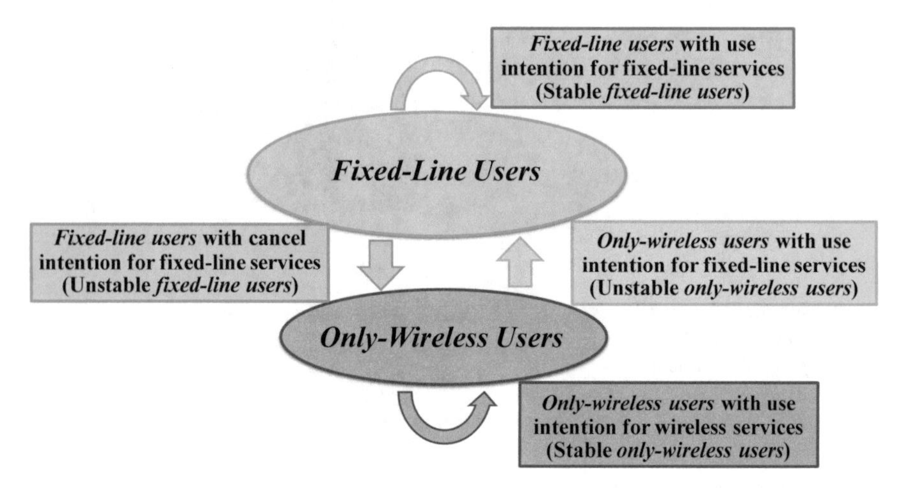

Fig. 4 User segments based on use intention for Internet access services

Table 5 Sample sizes by user segment of fixed-line users

User segment	Use intention	Number of samples	
Fixed-line users with use intention for fixed-line services (Stable *fixed-line users*)	I would like to use fixed-line service	724 (96.7%)	
Fixed-line users with cancel intention for fixed-line services (Unstable *fixed-line users*)	I may use only wireless service	17 (2.3%)	25 (3.3%)
	I will use only wireless service	8 (1.1%)	
All of *fixed-line users*	Total	749 (100%)	

Table 6 Sample sizes by user segment of only-wireless users

<table>
<tr><th>User segment</th><th>Use intention</th><th colspan="2">Number of samples</th></tr>
<tr><td rowspan="2">Only-wireless users with use intention for fixed-line services
(Unstable only-wireless users)</td><td>I may use both wireless service and fixed-line service</td><td>92
(21.0%)</td><td rowspan="2">153
(34.9%)</td></tr>
<tr><td>I may use only fixed-line service</td><td>61
(13.9%)</td></tr>
<tr><td>Only-wireless users with use intention for wireless services
(Stable only-wireless users)</td><td>I would like to use only wireless service</td><td colspan="2">286
(65.1%)</td></tr>
<tr><td>All of only-wireless users</td><td>Total</td><td colspan="2">439 (100%)</td></tr>
</table>

5 Internet-Access-Service Choice Behavior Analysis

5.1 Satisfaction Levels for Internet Access Services

We asked *fixed-line users* to evaluate the satisfaction level of the current fixed-line service against six factors. We also asked *only-wireless users* to evaluate the satisfaction level of current wireless service against seven factors. The satisfaction levels are defined by five levels: very satisfied, satisfied, nothing to complain, dissatisfied, very dissatisfied. Figures 5 and 6 present the percentage of respondents who are very satisfied and satisfied with the current internet access service. The results for *fixed-line users* and *only-wireless users* are shown in Fig. 5 and Fig. 6, respectively. Figures 7 and 8 present the percentage of respondents who are dissatisfied and very dissatisfied with the current service. The results for *fixed-line users* and *only-wireless users* are shown in Fig. 7 and Fig. 8, respectively.

It is found that there are several differences in the satisfaction levels between stable users and unstable users. The satisfaction level of stable *fixed-line users* against actual data transmission speed and stability is very high. The dissatisfaction level of unstable *fixed-line users* against monthly charge and discounts is higher than that of stable *fixed-line users*. The satisfaction level of stable *only-wireless* users against monthly charge is higher that of unstable *only-wireless* users. The dissatisfaction level of unstable *only-wireless users* against call quality is higher that of stable users.

5.2 Definition of Internet-Access Service Choice Model

We defined Internet access service choice model to analyze the decision-making factors in choices of Internet access services. This model is based on binominal logistic regression analysis. The formula is the same as that of binary logit discrete choice model [17, 18]. In this case, users can choose service 0 or service 1. The probability P_1 that a user will choose service 1 is given by

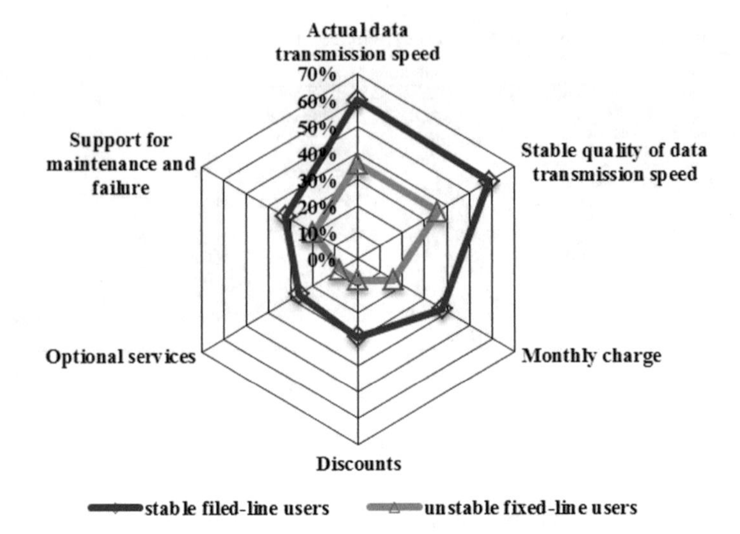

Fig. 5 Satisfaction levels of *fixed-line users*

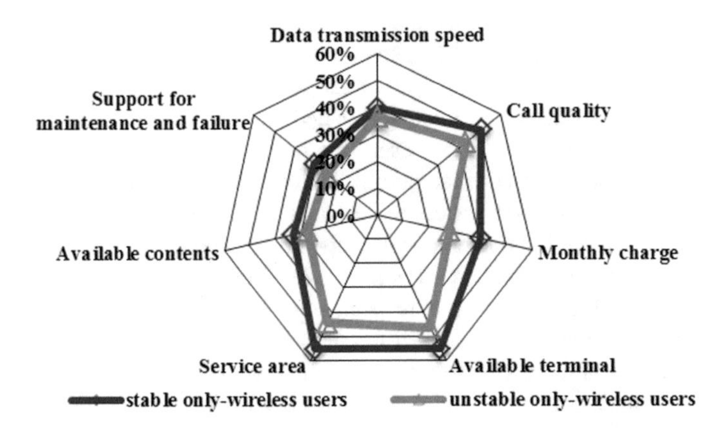

Fig. 6 Satisfaction levels of *only-wireless users*

$$P_1 = \frac{\exp(V_1)}{1 + \exp(V_1)} \tag{1}$$

V_1 is specified as follows:

$$V_1 = \alpha + \sum_{k=1}^{N} \beta_k x_k \tag{2}$$

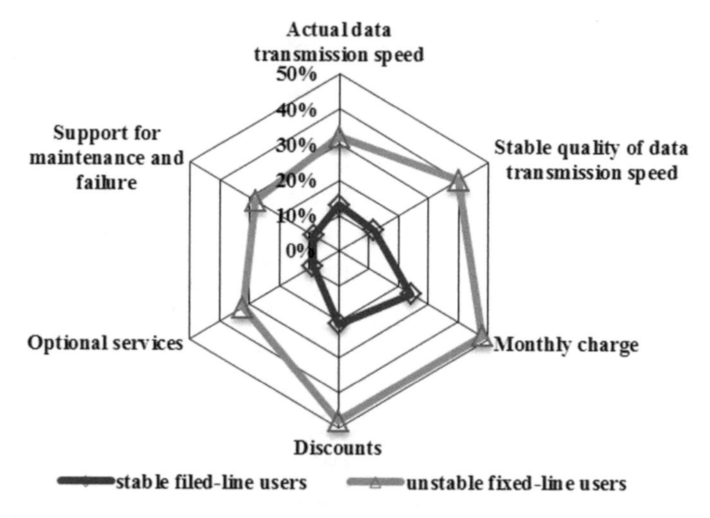

Fig. 7 Dissatisfaction levels of *fixed-line users*

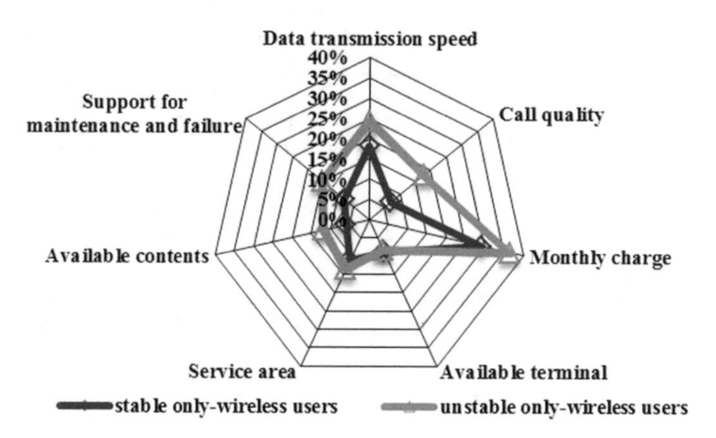

Fig. 8 Dissatisfaction levels of *only-wireless users*

where β_k ($k = 1, 2, \ldots, N$) is the coefficient denoting the weight for the explanatory value x_k of decision-making factor k, and α denotes the specific constant. The probability P_0 that a user will choose service 0 is calculated by $P_0 = 1 - P_1$.

5.3 Internet-Access Service Choice Model for Only-Wireless Users

We use the candidates for explanatory variables shown in Table 7. The coefficients of the explanatory variable x_k were estimated by using backward elimination method

Table 7 Definition of explanatory variables in choice model

Explanatory variables	Definition of variables
Number of Internet access users	This variable takes the value of 1 if the number of internet access users in same household is greater than 1, otherwise it takes the value of 0
Dissatisfaction with data transmission speed	This variable takes the value of 1 if a user's answer for data transmission speed is very dissatisfied or dissatisfied, otherwise it takes the value of 0
Dissatisfaction with call quality	This variable takes the value of 1 if a user's answer for call quality is very dissatisfied or dissatisfied, otherwise it takes the value of 0
Dissatisfaction with monthly charge	This variable takes the value of 1 if a user's answer for monthly charge is very dissatisfied or dissatisfied, otherwise it takes the value of 0
Dissatisfaction with available terminal	This variable takes the value of 1 if a user's answer for available terminal is very dissatisfied or dissatisfied, otherwise it takes the value of 0
Dissatisfaction with service area	This variable takes the value of 1 if a user's answer for service area is very dissatisfied or dissatisfied, otherwise it takes the value of 0
Dissatisfaction with available contents	This variable takes the value of 1 if a user's answer for available contents is very dissatisfied or dissatisfied, otherwise it takes the value of 0
Dissatisfaction with support for maintenance and failure	This variable takes the value of 1 if a user's answer for support for maintenance and failure is very dissatisfied or dissatisfied, otherwise it takes the value of 0

under the condition that p-value of each variable is less than 0.1. The estimation results are shown in Table 8 except the value of constant. The symbols: "**" and "*" indicate the significant levels of variables for the estimated model. The symbol "**" means that the p-value is less than 0.01. The symbol "*" means that the p-value is less than 0.05. We used an estimator, *BellCurve for Excel* by Social Survey Research Information Co., Ltd. [19] to estimate the coefficients of the model.

The estimated model can calculate the probability P_1 that a user will change to *fixed-line users*. This model also calculates the probability P_0 that a user will continue *only-wireless users*. The positive sign of coefficient indicates the reason why unstable *only-wireless users* will newly subscribe to fixed-line services.

The simulation results of two cases are shown in Table 8. If the value of each explanatory variable is the same as the set shown in Case I, the estimated probability P_1 becomes maximum. It means that if a user is dissatisfied with both monthly charge and call quality, and the number of Internet users in his/her household is greater than

Table 8 Estimated coefficients of choice model for only-wireless users

Explanatory variable x_k	Coefficients k	Value of explanatory variable x_k		p-value	
		Case I	Case II		
Dissatisfaction with available terminal	−0.9661	0	1	0.0352	*
Dissatisfaction with monthly charge	0.3908	1	0	0.0940	
Number of Internet access users	0.4404	1	0	0.0390	*
Dissatisfaction with call quality	1.3355	1	0	0.0004	**
Estimated propability of unstable *only-wireless users* : P_1		0.745	0.113		

*: P<0.05 **: P<0.01

1, the probability of newly subscribing to fixed-line services is about 0.745. If the value of each explanatory variable is the same as the set shown in Case II, the estimated probability P_1 becomes minimum. In other words, the probability P_0 becomes maximum. But, it is difficult to interpret the relationship between dissatisfaction with available terminal and increase in the probability P_0. Therefore we will explore the other explanatory variables and this is future study.

6 Supervised Learning Modelling

6.1 *Definition of Supervised Learning Model*

We divided stable *only-wireless users* into two segments by having experience of using fixed-line services or not. We also divided stable fixed-line users by the number of Internet users equals to one or not. Then, we analyzed the differences of these segments by using supervised learning model. This model is based on binominal logistic regression analysis. The probability P_1 that a user belongs to segment-1 is also given by Eq. 1 and V_1 is specified as Eq. 2. Tables 9 and 10 show the reasons for continuing current services, from which we chose the explanatory variables x_k in Eq. 2.

6.2 *Supervised Learning Models for Stable Only-Wireless Users*

The estimated model can classify stable *only-wireless users* into two segments: segment-0 (stable *only-wireless users* who had never used fixed-line services) and segment-1 (stable *only-wireless users* who had used fixed-line services before). The estimation results are shown in Table 11 except the value of constant. The factors

Table 9 Reasons for continuing current wireless services

Reasons for continuing current wireless services	Total of stable only-wireless users	Stable only-wireless users who had never userd fixed-line services	Stable only-wireless users who had used fixed-line services before.
I am satisfied with the data transmission speed of wireless services.	39.2%	35.6%	46.7%
I can use the home routers as the same usage as fixed-line services	33.2%	29.9%	40.2%
I would like to reduce communications costs.	23.8%	23.7%	23.9%
I usually use PC or Tablet with 3G or LTE service both in home and outside.	15.4%	12.4%	21.7%
Fixed-line telephpne service is not necessary.	14.7%	17.0%	9.8%
Cables and router to use fixed-line service are obstacle.	9.4%	9.8%	8.7%
The Interenet user is alone in my house.	9.1%	7.2%	13.0%
The low-priced and fixed-charge services which can transmit more than 20G packets are available.	8.7%	7.7%	10.9%
I can access the Internet by using tethering function of mobile phones.	7.7%	7.2%	8.7%
I get tired of contruct new fixed-line service.	7.7%	9.3%	4.3%
I use mobile PS or Tablet unable to connect by LAN cable in home.	7.0%	5.2%	10.9%
I think that the fixed-line service is unnecessary.	5.9%	7.7%	2.2%

of which the coefficients indicate negative express the characteristics of segment-0 users. The factor of which the coefficient indicates positive expresses the characteristics of segment-1 users. The estimated probability P_1 expresses the probability that a user belongs to the segment-1. The simulation results of two cases are shown in Table 11. If the value of each explanatory variable is the same as the set shown in Case I, the estimated probability P_1 becomes maximum. Therefore, segment-1 users tend to be single person households and consider data transmission speed of wireless services is same as that of fixed-line services.

Table 10 Reasons for continuing current fixed line services

Reasons for continuing current fixed-line services	Total of stable fixed-line users	Stable fixed-line users, the number of Internet users in household is greater than one	Stable fixed-line users, the number of Internet users in household is equal to one
I need data transmisssion speed of fixed-line services.	45.9%	46.3%	44.7%
I would like to use mobile terminals by WiFi router connected to fixed-line services.	28.5%	33.1%	16.1%
I need stable communications quality of fixed-line services.	25.1%	24.8%	26.1%
I would like to continue to use fixed-line telephpne service.	18.9%	17.9%	21.6%
I get tired of cancel the current service	15.5%	14.3%	18.6%
I would like to use fixed-charge service with no usage limitation.	14.8%	13.1%	19.1%
I have no reason to cancel fixed-line services.	14.4%	13.1%	17.6%
I can use fixed-line services very cheaply if I use bundle service with smart phones.	12.0%	14.5%	5.5%
I would like to use terminals via LAN cable.	11.3%	11.2%	11.6%
I feel uneasy that I use only wireless services.	9.5%	8.2%	13.1%
I need the stable environment that plural people can use at the same time from multiple terminals	8.6%	10.9%	2.5%
I would like to continue to use homepages and storage services provided by the curent provider.	7.6%	7.8%	7.0%
I am not burdened with the charge of fixed-line service.	6.8%	5.5%	10.1%

Table 11 Estimated coefficients of supervised learning model for stable only-wireless users

Explanatory Variable x_k	**Coefficients β_k**	**Value of Explanatory Variable x_k**		**p-value**	
		Case I	**Case II**		
I think that the fixed-line service is unnecessary.	-2.1524	0	1	0.0137	*
I am satisfied with the data transmission speed of wireless services.	0.6189	1	0	0.0237	*
I can use the home routers as the same usage as fixed-line services	0.6800	1	0	0.0170	*
I usually use PC or Tablet with 3G or LTE service both in home and outside.	0.8341	1	0	0.0204	*
The Interenet user is alone in my house.	1.3131	1	0	0.0071	**
Estimated propability that a user had used fixed-line services before: P_1 (Estimated probability that a user belongs to the segment-1)		0.884	0.028		

*: P<0.05 **: P<0.01

Table 12 Estimated coefficients of supervised learning model for stable fixed-line users

Explanatory Variable x_k	Coefficients β_k	Value of Explanatory Variable x_k		p-value	
		Case I	Case II		
I need the stable environment that plural people can use at the same time from multiple terminals	-1.9295	0	1	0.0002	**
I would like to use mobile terminals by WiFi router connected to fixed-line services.	-0.8960	0	1	0.0001	**
I can use fixed-line services very cheaply if I use bundle service with smart phones.	-0.8886	0	1	0.0097	**
I feel uneasy that I use only wireless services.	0.6454	1	0	0.0315	*
I would like to use fixed-charge service with no usage limitation.	0.7906	1	0	0.0028	**
I am not burdened with the charge of fixed-line service.	0.7915	1	0	0.0241	*
Estimated propability that the number of Internet users is equal to one: P_1 (Estimated probability that a user belongs to the segment-1)		0.805	0.020		

*: P<0.05 **: P<0.01

6.3 *Supervised Learning Model for Stable Fixed-Line Users*

The estimated model can classify stable *fixed-line users* into two segments: segment-0 (the number of internet users in household is greater than one) and segment-1 (the number of internet users is equal to one). The estimation results are shown in Table 12 except the value of constant. The factors of which the coefficients indicate negative express the characteristics of segment-0 users. The factors of which the coefficients indicate positive express the characteristics of segment-1 users. The estimated probability P_1 expresses the probability that a user belongs to the segment-1. The simulation results of two cases are shown in Table 12. If the value of each explanatory variable is the same as the set shown in Case I, the estimated probability P_1 becomes maximum. For segment-1 users, it is important to access internet with reliability and no usage limitation. If the value of each explanatory variable is the same as the set shown in Case II, the estimated probability P_0 becomes maximum. For segment-0 users, it is important that plural people can access the Internet comfortably and cheaply.

7 Conclusion

The purpose of this study is to understand the Internet access service choice behavior considering the current fixed-line and wireless service market in Japan. We have studied the differences between *fixed-line users* and *only-wireless users* from various viewpoints. It is found that there are two types of users: stable users and unstable users in these two user segments. In order to clarify the churning behavior between *fixed-line users* and *only-wireless users*, we focus on the following four customer-segments: stable *fixed-line users*, stable *only-wireless users*, unstable *fixed-line users* and unstable *only-wireless users*. We proposed one choice behavior model and two types of supervised learning models to create differential descriptions of these user segments. The choice model is constructed to clarify the differences between two segments: stable *only-wireless users* and unstable *only-wireless users*. It is found that dissatisfaction level had a big influence on user intention of choosing a service. Two types of supervised learning models are constructed to analyze stable *only-wireless users* and stable *fixed-line users*. It is found that the differences of reason to continue current services are related to user experience of the service that used before and user attributes. The analysis results in this paper have not yet been observable as a big movement of telecommunication market in Japan. But near future, the service using 5G technology will be started and differences between fixed-line services and wireless services will become small. We will consider these changes to understand the future market and analyze choice behavior according to new conditions.

References

1. Ministry of Internal Affairs and Communications, Information and Communications in Japan: White Paper 2018, Part 2, Chapter 5, Section 2 ICT Service Usage Trends. http://www.soumu.go.jp/johotsusintokei/whitepaper/eng/WP2018/chapter-5.pdf#page=11. Accessed Feb 2019
2. Ministry of Internal Affairs and Communications: Official Announcement of Quarterly Data on the Number of Telecommunications Service Subscriptions and Market Share (in Japanese). http://www.soumu.go.jp/main_content/000590807.pdf. Accessed Dec 2018
3. Inoue, A., Takahashi, S., Nishimatsu, K., Kawano, H.: Service demand analysis using multi-attribute learning mechanisms. In: 2003 IEEE International Conference on Integration of Knowledge Intensive Multi-agent Systems (KIMAS 2003), pp. 634–639 (2003)
4. Kurosawa, T., Inoue, A., Nishimatsu, K., Ben-Akiva, M., Bolduc, D.: Customer-choice behavior modeling with latent perceptual variables. Intell. Eng. Syst. Artif. Neural Netw. (ASME Press, NY) **15**, 419–426 (2005)
5. Nishimatsu, K., Inoue, A., Kurosawa, T.: Service-demand-forecasting method using multiple data sources. In: 12th International Telecommunications Network Strategy and Planning Symposium (NETWORKS2006), Technical Session 2.3 (2006)
6. Kurosawa, T., Inoue, A., Nishimatsu, K.: Service-choice behavior modeling with latent perceptual variables. Int. J. Electron. Cust. Relat. Manag. **2**(3), 228–250 (2008)
7. Takano, Y., Inoue, A., Kurosawa, T., Iwashita, M., Nishimatsu, K.: Customer segmentation in mobile carrier choice modeling. In: 9th IEEE/ACIS International Conference on Computer and Information Science (ICIS 2010), pp. 111–116 (2010)
8. Kurosawa, T., Bolduc, D., Ben-Akiva, M., Inoue, A., Nishimatsu, K., Iwashita, M.: Demand analysis by modeling choice of Internet access and IP telephony. Int. J. Inf. Syst. Serv. Sect. **3**(3), 1–26 (2011)
9. Inoue, A., Takano, Y., Kurosawa, T., Iwashita, M., Nishimatsu, K.: Mobile-carrier choice modeling framework under competitive conditions. J. Inf. Process. **20**(3), 585–591 (2012)
10. Inoue, A., Iwashita, M., Kurosawa, T., Nishimatsu, K.: Mobile-carrier choice behavior analysis around smart phone market. In: Proceedings of 14th IEEE/ACIS International Conference on Software Engineering, Artificial Intelligence, Networking and Parallel/Distributed Computing (SNPD2013), pp. 400–405 (2013)

11. Inoue, A., Tsuchiya, Y., Saito, M., Iwashita, M.: Demand analysis of Internet access services in Japan. In: 16th International Telecommunications Network Strategy and Planning Symposium (NETWORKS2014), Technical Session 14.2 (2014)
12. Inoue, A., Saito, M., Iwashita, M.: Behavior analysis on mobile-carrier choice & mobile-phone purchase. In: Proceedings of 2nd ACIS International Conference on Computational Science and Intelligence 2015 (CSI2015), CSI-SS3-1 (2015)
13. Inoue, A., Kitahara, K., Iwashita, M.: Behavior analysis on mobile-carrier choice considering mobile virtual network operators. In: Proceedings of 15th IEEE/ACIS International Conference on Computer and Information Science (ICIS 2016), pp. 995–1000 (2016)
14. Inoue, A., Kitahara, K., Iwashita, M.: Mobile-carrier choice behavior analysis between three major mobile-carriers and mobile virtual network operators. In: Proceedings of 18th IEEE/ACIS International Conference on Software Engineering, Artificial Intelligence, Networking and Parallel/Distributed Computing (SNPD2017), pp. 501–506 (2017)
15. Inoue, A., Satoh, A., Kitahara, K., Iwashita, M.: Mobile-carrier choice behavior analysis using supervised learning models. In: Proceedings of 7th International Congress on Advanced Applied Informatics (AAI 2018), pp. 829–834 (2018)
16. Provost, F., Fawcett, T.: Data Science for Business. O'Reilly Media, Inc., CA (2013)
17. Ben-Akiva, M., Lerman, S.R.: Discrete Choice Analysis. MIT Press, MA (1987)
18. McFadden, D.L.: The choice theory approach to market research. Mark. Sci. **5**(4), 275–297 (1986)
19. Social Survey Research Information Co., Ltd.: BellCurve for Excel. https://bellcurve.jp/ex/

Norm-referenced Criteria for Strength of the Upper Limbs for the Korean High School Baseball Players Using Computer Assisted Isokinetic Equipment

Su-Hyun Kim and Jin-Wook Lee

Abstract The purpose of this study is to set the norm-referenced criteria for isokinetic muscular strength of the upper limbs (elbow and shoulder joint) for the Korean 83 high school baseball players. HUMAC NORM (CSMI, USA) system was used to obtained the value of peak torque, peak torque per body weight. The results were presented as a norm-referenced criterion value using 5-point scale of cajori by 5 group (6.06, 24.17, 38.30, 24.17, and 6.06%). The provided criteria of peak torque and peak torque per body weight, set through the computer isokinetic equipment, are very useful information for high school baseball player, baseball coach, athletic trainer and sports injury rehabilitation specialists in injury recovery and return to rehabilitation, to utilize as an objective clinical assessment data.

Keywords Computer system · Baseball · Isokinetic equipment · Norm-referenced criteria · Elbow · Shoulder · Muscular strength

1 Introduction

In the current society of Information Industry, a computer not only helps the life of human more comfortable, also is being considered as a critical tool in the fields of space or aviation and is used in the wide and various range. It is also used in many areas in the physical exercise and sports, and its applicable scope is being extended.

A pitcher is a highly important position in a baseball which determines the issues of game by 80%, the role and position of pitcher are absolute as 68 players (61.8%)

S.-H. Kim (✉)
Department of Sports Medicine, Affiliation Sunsoochon Hospital, 76, Olympic-Ro, Songpa-gu, Seoul 05556, Republic of Korea
e-mail: trainerksh@hanmail.net

J.-W. Lee
Department of Exercise Prescription and Rehabilitation, Dankook University, 119, Dandae-Ro, Dongnam-gu, Cheonan-si, Chungcheongnam-do 31116, Republic of Korea
e-mail: rugby14@hanmail.net

R. Lee (ed.), *Big Data, Cloud Computing, and Data Science Engineering*, Studies in Computational Intelligence 844,
https://doi.org/10.1007/978-3-030-24405-7_8

out of 110 were pitchers in the first and second rookies draft for 2018 pro baseball [1].

Ball speed, control and consistency of speed are required to be a successful pitcher [2, 3], and it was reported that a pitcher who had in order of control, ability of power pitching, defense ability of long hits was more competitive accruing to Preceding Study on Performance Index based on data analysis of 2015 pro baseball [4].

40% of ball speed is contributed by step of lower body and movement of torso and rest is determined by shoulder, elbow and wrist joints, and the maximum spend can be generated at the upper limbs with the sufficient energy transfer from lower body [5, 6]. However, damages of shoulder or elbow joint may be occurred if the pitch for the high speed is performed when energy transfer is incomplete, so the balanced movement between upper and lower limbs when pitching is emphasized [7, 8].

It is reported that the core factor of baseball player is the instant react ability based on muscular strength [9], and the strong anaerobic power of upper limbs muscle [10, 11]. Regarding the physical factor related to muscle, the absolute muscle force is examined at stationary status, but it is well known that dynamic strength assessment is more effective since the muscular strength is generated through the movement, and the tension measurement should be done at the same time [12]. The isokinetic exercise, which causes muscular contraction at a constant speed, is performed under maximum load in all range of motion, and the tension occurred when the muscular contraction is effective for improvement of muscle strength since it accelerates the movement in the performance part. It is used to prevent injury and examine the damage of athletes or judge the progress of medical rehabilitation in the field of sports medicine [13]. It is also reported that the peak torque or peak torque per body weight which is the result of mechanical performance of muscle can be measured in a short period of time and it can be a critical data for the myofunctional examination that is highly reliable [14–16].

The muscular strength provides critical data for a coach or a trainer to set the training plan or assess the damages or rehabilitation progress [17]. High school baseball players have been greatly improved in terms of techniques and physical strength, but the objective assessing criteria for muscular strength are insufficient. In order to prevent injuries and improve performance for the baseball players, the objective assessing index for the lever of peak torque per body weight or training effect is required. Also, a study on the injury prevention and training for upper limbs is needed. On this study, the muscular strength of upper limbs of high school baseball players is measured in an objective and quantitative manners, and it may be provided for a coach or a trainer who set the training plan as a resource and can be criteria for return of injured players.

Table 1 The characteristics of subjects

N	Height (cm)	Weight (kg)	Age (years)	Career (years)	Body fat (%)
83	178.29 ± 6.04	78.07. ± 11.56	17.63 ± 0.69	6.53 ± 1.65	16.42 ± 5.97

Values are presented as mean ± standard deviation

2 Materials and Methods

Randomly selected 83 male baseball players of 10 high schools located in the Seoul, Gyeonggi-do area and registered with the Korea Baseball Association. The purpose of the study was thoroughly explained to the participants and consent was received that they will do their best. The physical characteristics of the subjects are as shown in Table 1.

2.1 Test Method for Extensor/Flexor of the Elbow Joint

HUMAC NORM (Stoughton, MA, USA), equipment for measuring isokinetic muscle strength was used to measure the muscles of the elbow. Calibration during testing was checked using the dynamometer's calibration program. All measurements were gravity compensated for limb weight. During elbow flexion/extension testing, the subjects were tested in a supine position with restraining straps placed across their shoulder and waist. The arm was positioned next to the torso at 0–10° of abduction. The forearm was pronated 90° such that the hand was in a neutral position. Pads were placed behind and lateral to the elbow to eliminate extraneous. The axis of rotation is immediately distal to the lateral epicondyle and moves only slightly anteriorly as flexion increase. The test was conducted 4 times at 60°/s [18]. For accurate measurement, a practice of 3 times was done at maximum muscle strength before the test, a rest was taken, and then the test was conducted (Fig. 1).

2.2 Test Method for External Rotator/Internal Rotator of the Shoulder Joint

HUMAC NORM (Stoughton, MA, USA), equipment for measuring isokinetic muscle strength was used to measure the muscles of the shoulder. The test was conducted four times at 60°/s. For accurate measurement, a practice of 3 times was done at maximum muscle strength before the test, a rest was taken, and then the test was conducted. The axis of rotation for shoulder internal/external rotation patterns is the longitudinal axis of the humerus. Position subject appropriately on chair, move and

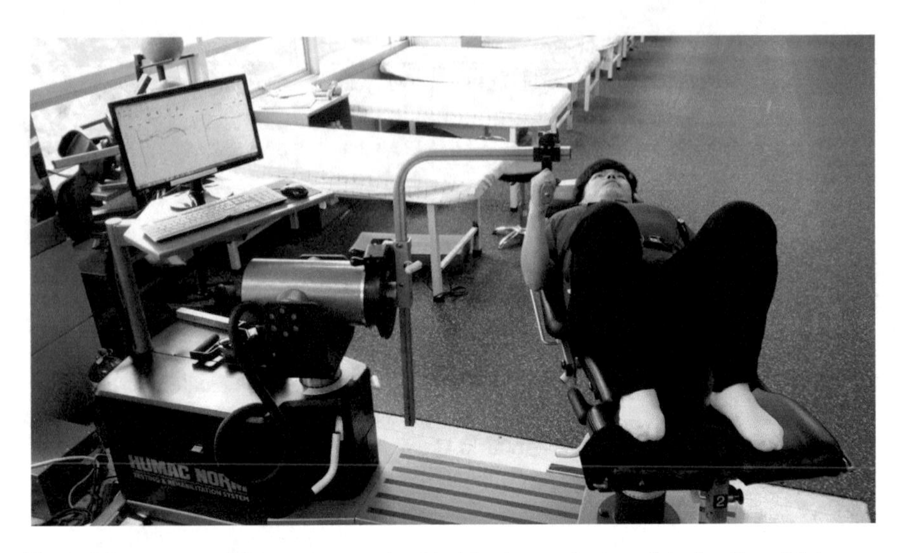

Fig. 1 Measurement of the computer assisted isokinetic muscle strength in the elbow joint

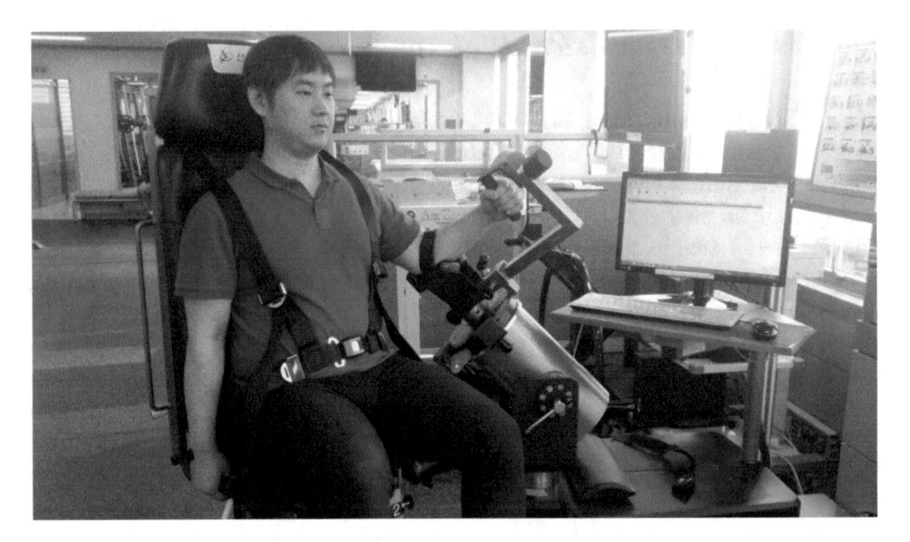

Fig. 2 Measurement of the computer assisted isokinetic muscle strength in the shoulder joint

secure chair at an appropriate distance from the dynamometer to permit subject to comfortably grasp adapter handgrip. Position subject's test arm with elbow flexed to 90° and shoulder slightly abducted. Stabilize forearm in elbow stabilizer pad. In order to minimize the involvement of other muscle groups or unnecessary movement, a belt was used to fix the chair (Fig. 2).

During the test, verbal encouragement was given to the subject to exert maximum muscle strength.

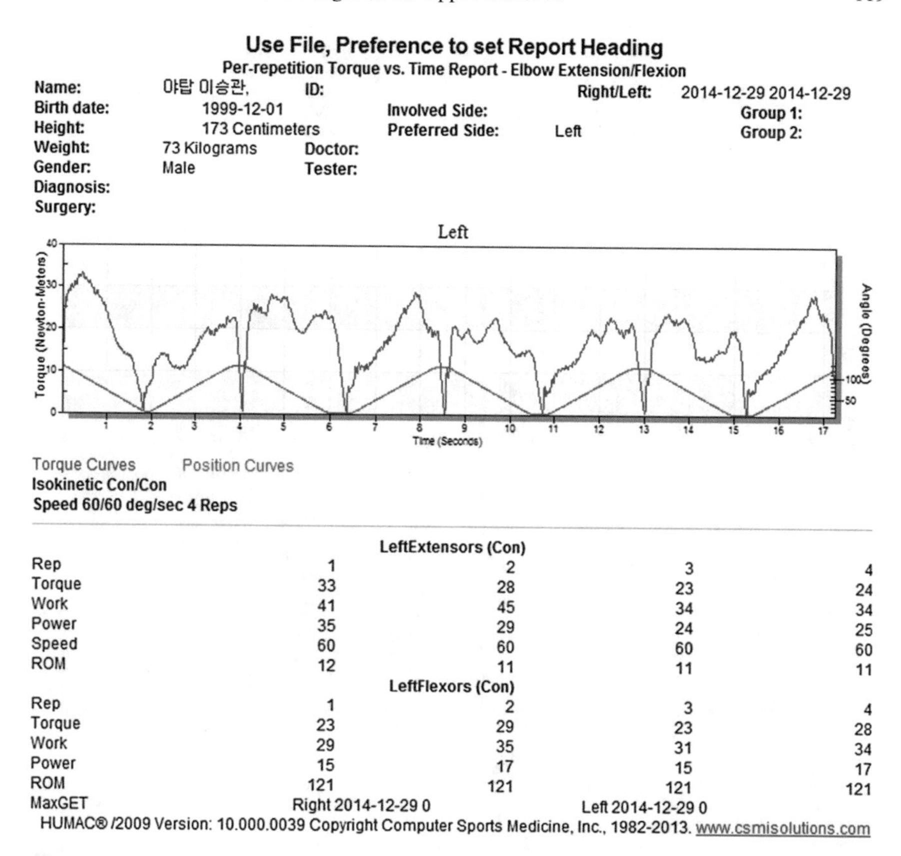

Use File, Preference to set Report Heading

Per-repetition Torque vs. Time Report - Elbow Extension/Flexion

Name: 아탑 이승관, **ID:** **Right/Left:** 2014-12-29 2014-12-29
Birth date: 1999-12-01 **Involved Side:** **Group 1:**
Height: 173 Centimeters **Preferred Side:** Left **Group 2:**
Weight: 73 Kilograms **Doctor:**
Gender: Male **Tester:**
Diagnosis:
Surgery:

Torque Curves Position Curves
Isokinetic Con/Con
Speed 60/60 deg/sec 4 Reps

	LeftExtensors (Con)			
Rep	1	2	3	4
Torque	33	28	23	24
Work	41	45	34	34
Power	35	29	24	25
Speed	60	60	60	60
ROM	12	11	11	11

	LeftFlexors (Con)			
Rep	1	2	3	4
Torque	23	29	23	28
Work	29	35	31	34
Power	15	17	15	17
ROM	121	121	121	121
MaxGET	Right 2014-12-29 0		Left 2014-12-29 0	

HUMAC® /2009 Version: 10.000.0039 Copyright Computer Sports Medicine, Inc., 1982-2013. www.csmisolutions.com

Fig. 3 Isokinetic torque recording: four successive repetition through the range of motion in elbow flexion/extension

Peak torque: The muscle strength measured without consideration given to body weight. The torque of the extensor and flexor after 4 repeats was measured at 60°/s. The unit is Nm [19]. The Isokinetic torque recording: four successive repetition through the range of motion in elbow flexion/extension is shown in Fig. 3.

Peak torque per body weight: The muscle strength measured with consideration given to body weight. The maximum torque of the extensor and flexor after 4 repeats is divided by the body weight. This figure represents relative muscle strength and the unit used is % [19]. The report of peak torque per body weight (%) in elbow flexion/extension is shown in Fig. 4.

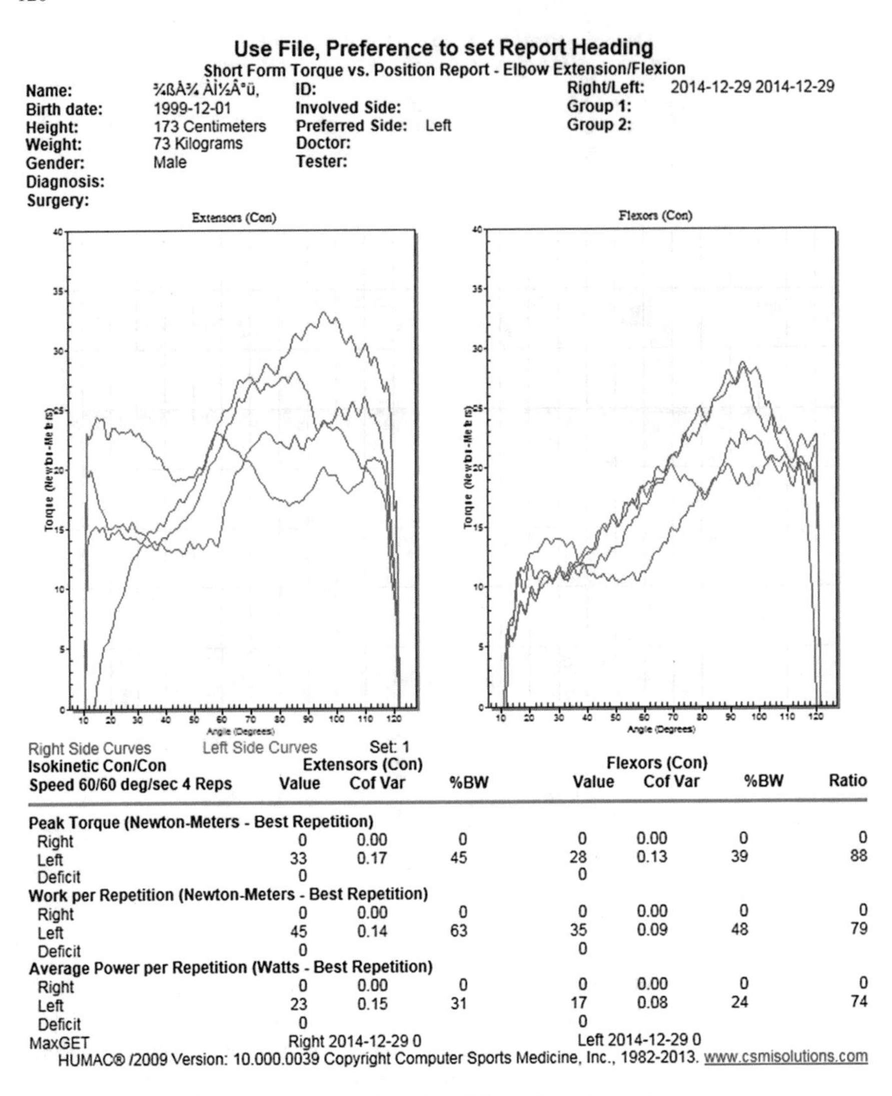

Use File, Preference to set Report Heading

Short Form Torque vs. Position Report - Elbow Extension/Flexion

Name:	¾ßÀ¾ ÀÌ½Â°ü,	**ID:**		**Right/Left:**	2014-12-29 2014-12-29
Birth date:	1999-12-01	**Involved Side:**		**Group 1:**	
Height:	173 Centimeters	**Preferred Side:**	Left	**Group 2:**	
Weight:	73 Kilograms	**Doctor:**			
Gender:	Male	**Tester:**			
Diagnosis:					
Surgery:					

Right Side Curves Left Side Curves Set: 1

Isokinetic Con/Con Speed 60/60 deg/sec 4 Reps	Extensors (Con) Value	Cof Var	%BW	Flexors (Con) Value	Cof Var	%BW	Ratio
Peak Torque (Newton-Meters - Best Repetition)							
Right	0	0.00	0	0	0.00	0	0
Left	33	0.17	45	28	0.13	39	88
Deficit	0			0			
Work per Repetition (Newton-Meters - Best Repetition)							
Right	0	0.00	0	0	0.00	0	0
Left	45	0.14	63	35	0.09	48	79
Deficit	0			0			
Average Power per Repetition (Watts - Best Repetition)							
Right	0	0.00	0	0	0.00	0	0
Left	23	0.15	31	17	0.08	24	74
Deficit	0			0			
MaxGET	Right 2014-12-29 0			Left 2014-12-29 0			

HUMAC® /2009 Version: 10.000.0039 Copyright Computer Sports Medicine, Inc., 1982-2013. www.csmisolutions.com

Fig. 4 The report of peak torque per body weight (%) in elbow flexion/extension

2.3 Data Analysis

The test was conducted four times at 60°/s. For accurate measurement, a practice of 3 times the collected data was analyzed thought SPSS 18.0 (SPSS Inc., Chicago, IL, USA) and the mean (M) and the standard error (SE) were calculated. The normal distribution was verified to set the criteria of the isokinetic muscular strength for elbow joint of high school baseball players. Given the variables, the results were applied percentile points considering the total number of cases and set by Cajori's

Table 2 Norm- referenced criterion value using 5-point scale of cajori

Scale	5	4	3	2	1
Percentage (%)	6.06	24.17	38.30	24.17	6.06
Stage	Very low	Low	Moderate	High	Very high

five-step internal estimation. The ratio of each range and criteria are as follows in Table 2.

3 Results and Discussion

3.1 Properties of the Sequence Selection

The analysis of measured variables using the HUMAC NORM (Stoughton, MA, USA) is as follows. The mean and standard deviation of the measured values for peak torque and peak torque per body weight in the extensor and flexor of the elbow joint at angular velocities of 60°/s are as seen in Table 3.

The mean and standard deviation of the measured values for peak torque and peak torque per body weight in the internal/external rotation of the shoulder joint at angular velocities of 60°/s are as seen in Table 4.

Table 3 The values of peak torque and peak per body weight elbow joint flexor and extensor at 60°/s speed

Speed	Muscle GRP	Side	Peak torque (Nm)	Peak torque per body weight (%)
60°/s	Fle	Dominate	38.25 ± 8.83	49.59 ± 8.99
	Ext	Dominate	46.64 ± 12.39	60.30 ± 12.87

Values are presented as mean ± standard deviation/Fle: flexor, Ext: extensor/GRP: group

Table 4 The values of peak torque and peak torque per body weight shoulder joint external rotator and internal rotator at 60°/s speed

Speed	Muscle GRP	Side	Peak torque (Nm)	Peak torque per body weight (%)
60°/s	IR	Dominate	48.73 ± 9.47	63.43 ± 11.23
	ER	Dominate	20.31 ± 5.82	26.39 ± 7.32

Values are presented as mean ± standard deviation/IR: internal rotator, ER: external rotator/GRP: group

Table 5 Norm-referenced criterion value of peak torque and peak torque per body weight elbow joint flexor and extensor at 60°/s speed

Speed	Muscle GRP	Stage	Very low	Low	Moderate	High	Very high
60°/s (PT)	Fle	Dominate	25.0 less	25.1–33.8	33.9–42.7	42.8–51.4	51.5 more
	Ext	Dominate	28.0 less	28.1–40.5	40.6–53.0	53.1–65.4	65.5 more
60°/s (PT%BW)	Fle	Dominate	37.1 less	37.2–45.7	45.8–54.3	54.4–63.0	63.1 more
	Ext	Dominate	39.0 less	41.0–53.9	54.0–66.9	67.0–79.9	80.0 more

PT peak torque(Nm), PT%BW peak torque per body weight (%)/Fle: flexor, Ext: extensor

Table 6 Norm-referenced criterion value of peak torque and peak torque per body weight shoulder joint internal rotator and external rotator at 60°/s speed

Speed	Muscle GRP	Stage	Very low	Low	Moderate	High	Very high
60°/s (PT)	IR	Dominate	34.5 less	34.6–44.0	44.1–53.5	53.6–62.9	63.0 more
	ER	Dominate	11.6 less	11.7–17.4	17.5–23.2	23.3–29.0	29.1 more
60°/s (PT%BW)	IR	Dominate	46.6 less	46.7–57.8	57.9–69.0	69.1–80.3	80.4 more
	ER	Dominate	15.4 less	15.5–22.7	22.8–30.1	30.2–37.4	37.5 more

PT: peak torque(Nm), PT%BW: peak torque per body weight(%)/Fle: flexor, Ext: extensor

The assessment reference value of the peak torque and peak torque per body weight for the flexor and extensor of the elbow joint at angular velocities of 60°/s are as shown in Table 5.

The assessment reference value of the peak torque and peak torque per body weight for the internal/external rotator of the shoulder at angular velocities of 60°/s are as shown in Table 6.

Damages in upper limbs were frequently found among baseball players due to the overuse, and if a high school player gets damages, it may cause the disabilities of joint or shorten the career because of the permanent transform.

It is reported that dissatisfaction of pitching style, pitching with fatigue shoulder, unbalanced muscular strength, increase in the number of pitches are the common causes [20, 21] and USA Baseball Medial and Safety Advisory Committee recommends that young players should restrict the number of pitches to prevent the common damages [22]. Yoon [23] who studied anaerobic power capacity and speed of pitching stated that increase in muscular strength improved the ball speed, and Pottriger et al. [11] reported that there was correlation between the muscular strength and the ball speed. For the reason, the importance of muscular strength is being emphasized. It was said that examining the muscular strength and pitch speed and applying to the result of game was extremely important [23], and in order to prevent injuries and improve performance for the baseball players, the objective assessing index for the lever of peak torque per body weight or training effect is required.

Kim et al. [25] researched the peak torque of flexor and extensor of elbow joint of 18 university pitchers and their averages were 55.17 ± 14.71 Nm and 53.50 ± 10.30 Nm respectively and the peak torque per body weight of them were reported as 55.83 ± 9.32 Nm and 59.67 ± 12.68 Nm. Jang [26] targeted high school baseball players and reported that the peak torque of flexor and extensor of elbow joint were averaged 40.94 ± 1.45 Nm and 50.13 ± 1.24 Nm. In the study, peak torque per body weight of flexor and extensor were 38.25 ± 8.83 Nm and 46.64 ± 12.39 Nm which were lower than university players, but the similar result showed when the university players were targeted. According to those studies, high school baseball players show lower lever of both peak torque and peak torque per body weight than university baseball players.

When pitching is performed, the external rotation is increased and the muscle of internal torque gets grater. It accelerates the ball speed and the wider angle of external rotation is at the beginning of pitch, the faster ball speed is [27] However, the internal rotation of shoulder joint decreases after pitch but the external rotation increases dramatically so that the balance of the rotator cuff, stationary structure of shoulder joint and its surrounding muscles may be broken Yanagisawa et al. [27].

In the study of Ma [24] that researched physique and muscular strength between university and pro baseball players showed that the averages of peak torque of shoulder internal rotation were 50.50 ± 11.40 Nm and 50.00 ± 6.5 Nm at 60°/s, and the averages of peak torque of external rotation were 44.70 ± 10.20 Nm and 42.60 ± 6.60 Nm. The peak torque per body weight of shoulder internal rotation averaged 61.20 ± 11.30 Nm and 50.20 ± 4.20 Nm respectively and it was reported that the difference was not statistically significant.

In the study of Jang [26] which was about the correlation between the muscular strength and pitch speed of high school pitchers indicated that the peak torque of shoulder internal rotation at 60°/s was 51.71 ± 1.44 Nm and the peak torque of external rotation was 22.21 ± 0.92 Nm which showed correlation with the pitch speed. In this study with similar target, the peak torque of shoulder internal rotation at 60°/s averaged 48.73 ± 9.47 Nm and the peak torque of external rotation was 63.43 ± 11.2 Nm. Also, the peak torque per body weight of shoulder internal rotation was 63.43 ± 11.2 Nm and the peak torque per body weight of external rotation was 26.29 ± 7.32 Nm. The peak torque and the peak torque per body weight were similar between pro and university baseball players. It indicates that high school baseball players improved in terms of physique but, the peak torque and the peak torque per body weight of external rotation were lower. Since the external rotation affects on the process of shoulder deceleration, increase in external rotation of shoulder joint through training may prevent damages, and the ratio of internal and external rotator should be 65–35% for the balance of muscular strength [28, 29]. Unbalanced muscular strength may cause the damages so that muscle strengthening exercise for the external rotator is required, and it should be carried out in a systematic program in order for high school pitchers to improve their muscular strength and performance. This study provides the quantitative and objective criteria for the muscular strength of shoulder and elbow joint for high school pitchers, it may be used as resources for a coach or a trainer who set the training plan, also may be the basic data when injured players recover or return by examining the muscular strength of shoulder and elbow joint of them. Succeeding studies should be conducted in various aspects to prevent injuries or damages of shoulder or elbow joint and a study targeted at primary and middle school baseball players should be carried out. This study is considered as high useful resource for not only baseball players also a coach or a trainer and suggests the precise criteria of muscular strength of shoulder and elbow joint for high school baseball player.

4 Conclusion

The purpose of this study is to provide the objective assessing criteria with a baseball coach or a rehabilitation professional who set the training plan and assess it. This study provides the quantitative and objective criteria for the muscular strength of upper limbs for high school pitchers, it may be used as resources for a coach or a trainer who set the training plan, also may be the basic data when an injured player recovers the damages or returns by examining the muscular strength of upper limbs of them. Succeeding studies should be conducted in various aspects to prevent injuries or damages of upper limbs and a study targeted at primary and middle school baseball players should be carried out as well. This study, which 83 high school pitchers were targeted, is considered as highly useful resource for a coach or a trainer in the field and suggests the precise criteria of muscular strength of upper limbs for high school baseball player.

References

1. Korea Baseball Softball Association (2017). http://www.koreabaseball.com/News/Notice/View.aspx?bdSe=6780
2. Stodden, D.F., Fleisig, G.S., McLean, S.P., Lyman, S.L., Andrews, J.R.: Relationship of pelvis and upper torso kinematics to pitched baseball velocity, J. Appl. Biomech. **17**(2), 164–172 (2001)
3. Matsuo, T., Escamilla, R.F., Fleisig, G., Barrentine, S.W., Andrews, J.R.: Comparison of kinematic and temporal parameters between different pitch velocity groups. J. Appl. Biomech. **17**(1), 1–13 (2001)

4. Lee, S.: Development of Pitcher's Performance Index in the Korean Professional Baseball Games. Graduate School, Myongji University (2015)
5. Hoshikawa, T., Toyoshima, S.: Contribution of body segments to ball velocity during throwing with non-preferred hand. Biomechanics VB. 109–117 (1976)
6. Son, H.H.: A Three–Dimension al Kinematic Analysis of the Upper Body During the Straight—Ball Pitches. Graduate School, Yonsei University (2001)
7. Dun, S., Loftice, J., Fleisig, G.S., Kingsley, D., Andrews, J.R.: A biomechanical comparison of youth baseball pitches: is the curveball potentially harmful. Am. J. Sports Med. **36**(4), 686–692 (2008)
8. Lim, D.J.: Kinematic Characteristics of the upper limb segment of College baseball pitchers pitching a fastball and curve ball. Graduate School, Konkuk University (2013)
9. Oh, D.J., Park, B.J.: The study on the relationship between baseball players batting average and physical strength. J. Korean Soc. Study Phys. Educat. **7**(4), 241–254 (2003)
10. Cho, J.H., Lim, S.K., Kwon, T.Y., Kim, B.G., Ahn, Y.W.: Clinical article: the correlation between throwing speed and shoulder internal/external rotator, trunk flexor/extensor, knee flexor/extensor strength and power in the professional baseball pitchers. Korean Soc. Sports Med. **24**(2), 158–163 (2006)
11. Potteiger, J.A., Wilson, G.D.: Bridging the gap-research: training the pitcher: a physiological perspective. Strength Condition J. **11**(3), 24–26 (1989)
12. Humer, M., Kösters, A., Mueller, E.: Dynamic and static maximum strength in closed kinetic chain movements, Trunk flexion/-extension and-rotation. Sportverletzung Sports chaden: Organ der Gesellschaft fur Orthopadisch-Traumatologische Sportmedizin, **25**(1), 13–21 (2011)
13. Sherman, W.M., Plyley, M.J., Vogelgesang, D., Costill, D.L., Habansky, A.J.: Isokinetic strength during rehabilitation following arthrotomy: Specificity of speed. Athletic Training. **16**(1), 138–141 (1981)
14. Beam, W.C., Bartels, R.L., Ward, R.W., Clark, R.N., Zuelzer, W.A.: Multiple comparisons of isokinetic leg strength in male and female collegiate athletic teams. Med. Sci. Sports Exerc. **17**(2), 269 (1985)
15. Gene, M.A.: Exercise Physiology: Laboratory Manual, pp. 178–183. William C Brown Publisher (1998)
16. Perrine, H.D.: Isokinetic Exercise and Assessment. Human Kinetics Publisher (1994)
17. Jones, P.A., Bampouras, T.M.: A comparison of isokinetic and functional methods of assessing bilateral strength imbalance. J Strength Condition Res **24**(6), 1553–1558 (2010)
18. Kim, S.H.: The effects of physique and physical fitness on ball speed in high school baseball players. Dankook University, Graduate School (2015)
19. Kim, S.H., Yoon, W.Y.: A study on the Norm-referenced criteria for isokinetic functional strength of the wrist for junior baseball players. Indian J. Sci. Technol. **8**(18), IPL071 (2015)
20. Lyman, S., Fleisig, G.S., Waterbor, J.W., Funkhouser, E.M., Pulley, L., Andrews, J.R., Roseman, J.M.: Longitudinal study of elbow and shoulder pain in youth baseball pitchers. Med. Sci. Sports Exerc. **33**(11), 1803–1810 (2001)
21. Tullos, H.S.: Throwing mechanism in sports. Orthop. Clin. North Am. **4**(3), 709–720 (1973)
22. USA Baseball Medical and Safety Advisory Committee, "Position statement on youth baseball injuries." www.usabaseball.com/med_position_statement.html. Accessed 07 Nov 2004
23. Yoon, Jeong. Hyun., "Relationship between anaerobic power and pitching velocity of baseball athletes", Korean journal of physical eduaction., **34**(3), 3270–3275 (1995)
24. Ma, H.Y.: Comparison of Body Composition, Physical Fitness, and Isokinetic Strength Between Pro-Baseball Players and College Baseball Players. Graduate School, Korea University (2010)
25. Kim, Y.B., Kim, J.H., Chol, J.I., Cho, W.J., Shin, J.Y.: The correlation fast ball and slider speed physique related isokinetic muscular peak torque of university baseball pitchers. Korean J. Sports Sci. **23**(1), 1123–1133 (2014)
26. Jang, JS.: The Correlation of Ball Speed and Isokinetic Muscular Strength in High-School Baseball Pitchers. Graduate School, Dankook University (2015)
27. Yanagisawa, O., Miyanaga, Y., Shiraki, H., Shimojo, H.: The effects of various therapeutic measures on shoulder strength and muscle soreness after baseball pitching. J. Sports Med. Phys. Fitness **43**(2), 189 (2003)

28. Alderink, G.J., Kuck, D.J.: Isokinetic shoulder strength of high school and college-aged pitchers. J. Orthop. Sports Phys. Ther. **7**(4), 163–172 (1986)
29. Wilk, K.E., Arrigo, C.A., Andrews, J.R.: Current concepts: the stabilizing structures of the glenohumeral joint. J. Orthop. Sports Phys. Ther. **25**(6), 364–379 (1997)

A Feature Point Extraction and Comparison Method Through Representative Frame Extraction and Distortion Correction for 360° Realistic Contents

Byeongchan Park, Youngmo Kim and Seok-Yoon Kim

Abstract 360° realistic contents are omnidirectional media contents that support front, back, left, right, top and bottom. In addition, they are combined images of images produced using two or more cameras through the stitching process. Therefore 4K UHD is basically supported to represent all directions and distortion occurs in each direction, especially above and below. In this paper, we propose a feature point extraction and similarity comparison method for 360° realistic images by extracting representative frames and correcting distortions. In the proposed method, distortion-less parts for an extracted frame such as the front, back, left, and right directions of the image, except for the largest distortion area such as the up and down directions, are first corrected by a rectangular coordinate system method. Then, the sequence for the similar frames is classified and the representative frame is selected. The feature points are extracted from the selected representative frames by the distortion correction and the similarity can be compared in the subsequent query images. The proposed method is shown, through the experiments, to be superior in speed for the image comparison than other methods, and it is also advantageous when the data to be stored in the server increase in the future.

Keywords 360° realistic contents · Distortion correction · Sequence classification · Feature-point extraction · Similarity comparison

B. Park · Y. Kim (✉) · S.-Y. Kim
Department of Computer Science and Engineering, Soongsil University, Seoul, Republic of Korea
e-mail: ymkim828@ssu.ac.kr

B. Park
e-mail: pbc866@ssu.ac.kr

S.-Y. Kim
e-mail: ksy@ssu.ac.kr

R. Lee (ed.), *Big Data, Cloud Computing, and Data Science Engineering*, Studies in Computational Intelligence 844,
https://doi.org/10.1007/978-3-030-24405-7_9

1 Introduction

Recently, as interests in Virtual Reality (VR) increase in the 4th industrial revolution era, much research has been conducted on it [1]. VR technology is a technology that makes it possible to use a virtual environment, and it is an interface between a human and a computer that allows an indirect experience of a real environment by interacting with a human. There are two main types of VR: VR contents based on 3D animation using computer graphic technology and VR contents created by realizing virtual reality using a computing technology by photographing actual places with a camera [2–4].

According to TrendForce, global VR market has an annual average growth rate (CAGR) of 77.8% from $ 6.7 billion as of 2016, and it will grow exponentially in billions of dollars. In addition, the market has changed from VR hardware-oriented distribution to content-oriented distribution starting from 2018. The content applications to be applied to VR technology include games, education, medical, broadcasting, advertising, and video, etc.

Recently, companies have developed 360° VR cameras and Head Mount Displays (HMD) which can easily use the VR technology to make and use 360° VR realistic contents, and various low-priced and high-end products that can be tailored to the user's environment have been released, making it possible to create and use VR contents easily for both industries and individuals.

In accordance with this growing market environment, several standardization groups such as MPEG, 3GPP, and DVB have announced standardization of VR technology. Especially, MPEG in December, 2017, proposed MPEG-I (MPEG-Immersive) which contains the formats supporting immersive and omni-directional videos and streaming technology, and has finally completed VR standardization work supporting 6-DoF (Degree of Freedom) [5, 6].

However, contrary to the development of VR industry and technology, VR contents happen to be illegally distributed through torrents and web hards, etc., by transforming contents or dissolving copyright protection mechanism such as DRM (Digital Right Management). Most VR contents require a high production cost, so copyright should be protected. Since the VR contents are reproduced from a plurality of frames as in the case of normal 2D contents, distortion correction, feature point extraction, and similarity comparison of all the frames of a high-capacity and high-quality VR content are much difficult [7–14].

In this paper, we first extract a representative frame from one sequence composed by a consecutive frames for all extracted frames, to expedite feature extraction and similarity comparison in VR contents. Since the feature points are more precisely extracted from the frames through the distortion correction than from the frames without the distortion correction, the distortion of the representative frame of the extracted sequence is corrected in the directions other than the top and bottom where the distortion is most severe among all directions. The ORB algorithm is used to extract the feature points for selecting the representative frames by specifying the sequence. The most accurate method is selected by comparing the similarity between the frame with distortion correction and the frame without distortion correction.

The composition of this paper is as follows. Section 2 explains the production process and characteristics of VR contents, the algorithm of image feature points extraction and matching process. Section 3 shows the characteristics of VR contents and ERP, and proposes the feature point extraction method through distortion correction of spherical coordinate system and frame sequence classification method for fast similarity comparison. In Sect. 4, the validity of the proposed method is verified through performance evaluation. Section 5 concludes the paper.

2 Related Research

2.1 VR Contents Production and Characteristics

MPEG defined OMAF (Omnidirectional Media Format) as the production and use process of VR contents through standardization work of MPEG-I. VR contents are created using VR camera and technical part is illustrated in Fig. 1, so that contents can be used by HMD.

During the process of VR contents being produced, VR contents will be filmed with two or more cameras, stitching and projection into 3-D Rectangle images, and users will use VR contents through HMD. In particular, the process of expressing

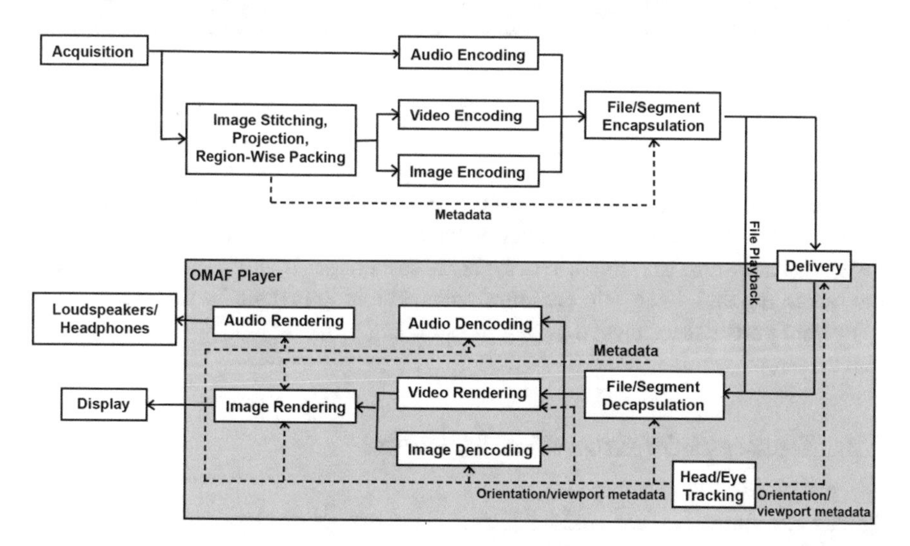

Fig. 1 OMAF architecture of MPEG-I

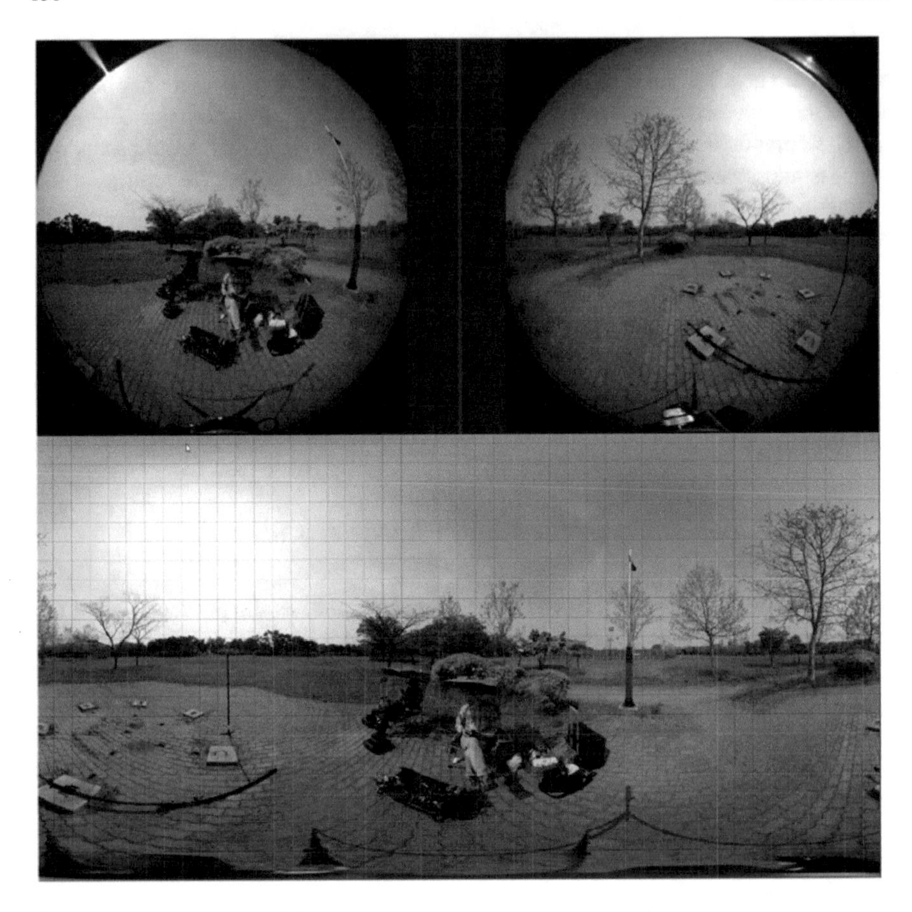

Fig. 2 Image extraction and ERP

3D spherical images as 2D images is called projection, and the most commonly used method for VR content is ERP (Equi-Rectangular Projection), as shown in Fig. 2.

In VR contents, these ERP methods have distorted areas because they contain forward, backward, up, down, left, right in one image. In addition, some parts of the image (up and down) are unnatural when stitching is done by forcing the non-redundant parts to be attached during the stitching process.

2.2 Feature Point Extraction Algorithm

Three main algorithms are used for feature-point extraction algorithms [15–18]. The first SIFT (Scale Invariant Feature Transform) algorithm was proposed by Lowe David in 2004 to address issues sensitive to changes in image scale. As a result, it is

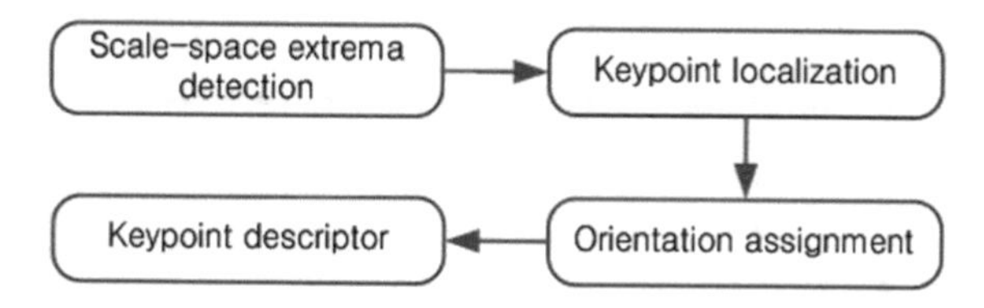

Fig. 3 SIFT algorithm

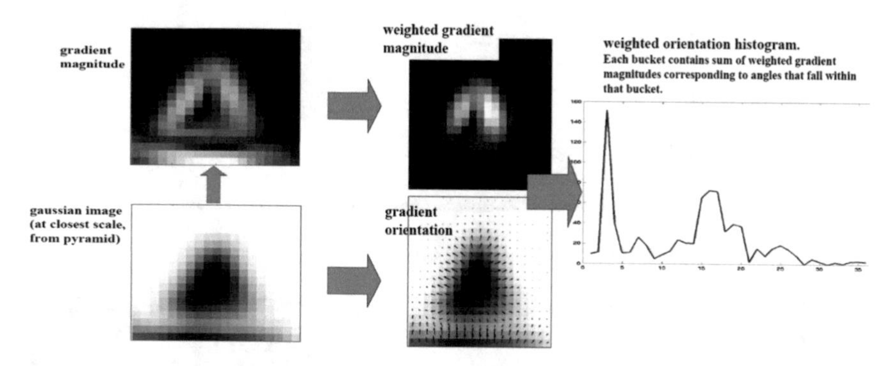

Fig. 4 Directional assignment process are extracted by the gradient direction and size

tough not only to change the scales, but also to change the image rotation, similarity variation, perspective change, noise and lighting. Figure 3 shows the implementation of the SIFT algorithm [11, 19].

In the first step, the detection of the scale-space pole value, an image pyramid is generated, the value of Difference of Gaussian (DoG) is obtained for each scale image, and the extremum is used to identify the position and scale of the principal points. In the second step of major point localization, the Taylor series is used to find the pole and the point value to remove candidate feature points. The third step is a step of allocating the main directions of the extracted feature points, and a histogram is obtained from the inclination directions of the sample points in the peripheral region of the point for direction assignment (Fig. 4).

In the last step, the directions of the directional assignment are extracted by extracting the direction and size of the tilt. SIFT finds the direction of the tilt values around the main point. The slope value is weighted using a Gaussian weight function, and the size of the slope value is calculated by using a value corresponding to 1.5 times of the corresponding key point scale value.

The second SURF (Speed Up Robust Features) algorithm is proposed by Bay in 2008 as a multi-scale space theory based algorithm. The feature descriptor is a Hessian matrix with excellent performance and accuracy. The SURF algorithm shows the same performance as the SIFT algorithm, but it is a faster algorithm than the SIFT algorithm. It uses a box filter and an integral image for speedup. The flow of the

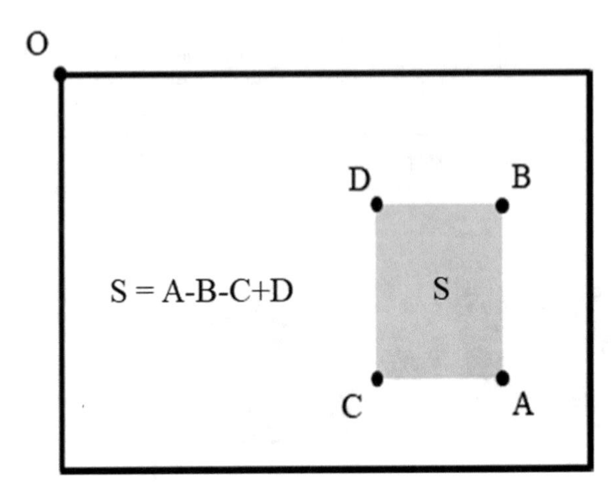

Fig. 5 Generation of integral image

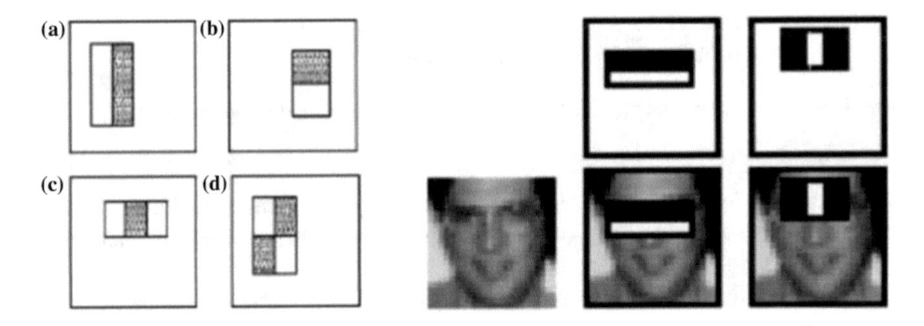

Fig. 6 Haar-like feature detection method

algorithm proceeds with the feature point extraction, main direction determination, and descriptor generation similar to the SIFT algorithm [20, 21] (Fig. 5).

Finally, the ORB (Oriented FAST and Rotated BRIEF) algorithm is a mixture of FAST and BRIEF. It is an image feature point detection algorithm developed by OpenCV LAB to replace commercial SIFT and SURF due to licensing problems. The basic idea of the ORB is to find the feature points using the FAST algorithm and apply the Harris corner detection method to compute the top N of the feature points found. In addition, various scaled pyramids are applied to extract size and invariant image features, and the adjusted BRIEF algorithm is applied to extract rotation invariant features as well.

The Haar-Like Feature is an algorithm proposed by Viola in 2001 [22]. As shown in Fig. 6, it uses four rectangular detection methods of A and B (two-rectangle feature), C (three-rectangle feature) and D (four-rectangle feature), and the brightness value of the region is subtracted from the image to find the value bigger than the threshold. In addition, it generates an integral image and obtains a value regardless of its size and position by total four additions.

3 Feature Point Extraction and Similarity Comparison Method Using Distortion Correction

3.1 Frame Extraction and Distortion Correction

The number of frames for one hour of video content is about 100,000–120,000, depending on the image quality. The higher the frame, the smoother and clearer the video content. VR content is often captured by fixing a 360° camera at a specific place, so similar scenes continue even if the number of frames is large.

Also, VR contents such as Fig. 4 are video clips that are uploaded through video editing after filming, with each location to be filmed. Thus, a series of similar frames can be classified within the image content into each sequence representing the start and end of K of a particular frame (Fig. 7).

The ERP method is generally used to express VR content as a monitor. The projection method of ERP presents all scenes (Front, Back, Right, Left, Top, Bottom) taken with a 360° camera as one image (Fig. 8).

In the ERP method, the front, back, left, and right except the top and bottom are not relatively distorted. Therefore, the image projected by ERP is divided into 3 parts by Top, Middle and Bottom, and the Middle part excluding the Top and Bottom with the worst distortion is divided into 4 parts and distortion correction is performed by the spherical coordinate system (Fig. 9).

Distortion correction is performed by dividing four regions in the middle region and uses a spherical coordinate system (Fig. 10).

3.2 Feature Point Extraction and Sequence Classification

In this paper, feature points are extracted by using ORB algorithm for frames whose distortion correction is completed by specifying four regions in one frame. Feature points are extracted using the ORB algorithm in the specified four regions. Since the ORB algorithm compares feature points extracted at the matching step, although the feature point extraction is fast, it has a disadvantage that the execution speed becomes slower as the number of images and feature points increases. In order to expedite similarity comparison for later image query, ORB algorithm is used to group similar frames and to classify sequences and select representative frames to make similarity comparison.

A frame is extracted from the VR contents to designate an area and perform distortion correction. The feature points are extracted by the ORB algorithm in the

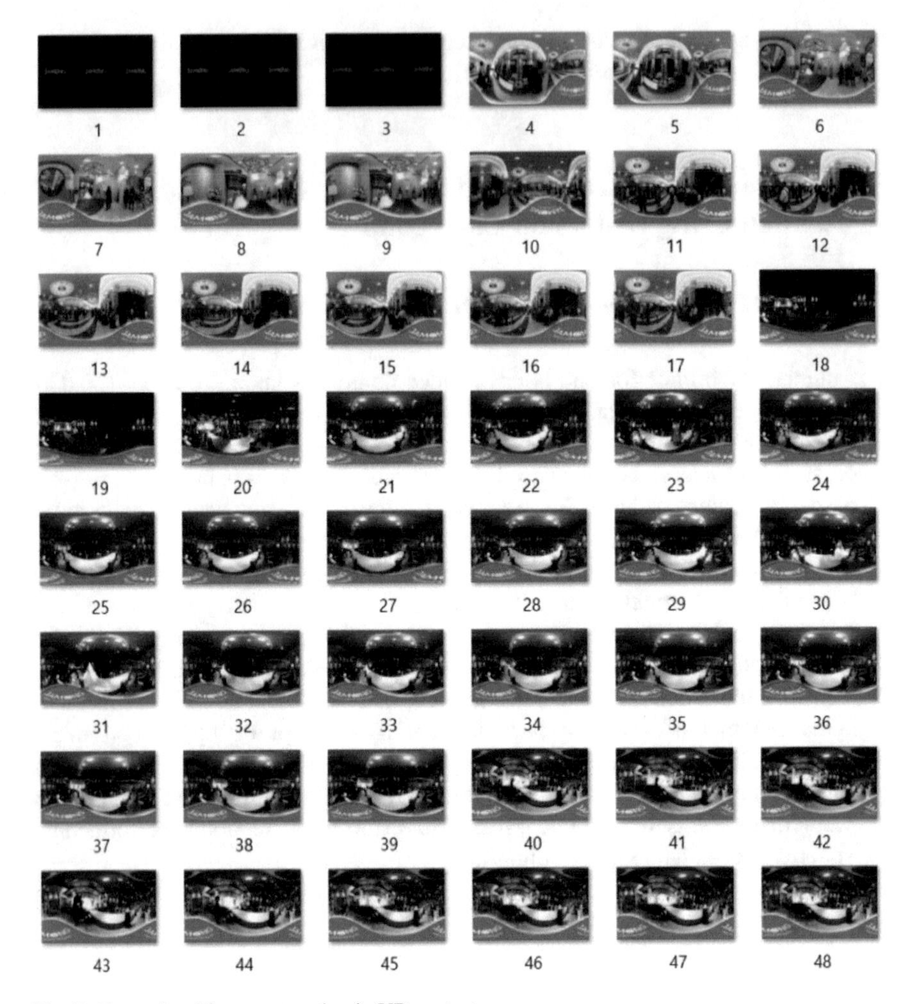

Fig. 7 Example of frame extraction in VR content

region where the distortion is corrected. The feature points extracted from each frame are compared and the similarity is compared. In the comparison process, matching numbers are designated, and when a matching number equal to or greater than a certain number is found, the similar frames are judged to be classified into one sequence. If a matching number equal to or greater than a predetermined value is not found, the frame is judged to be a different frame and classified into another sequence.

Fig. 8 Expression of ERP

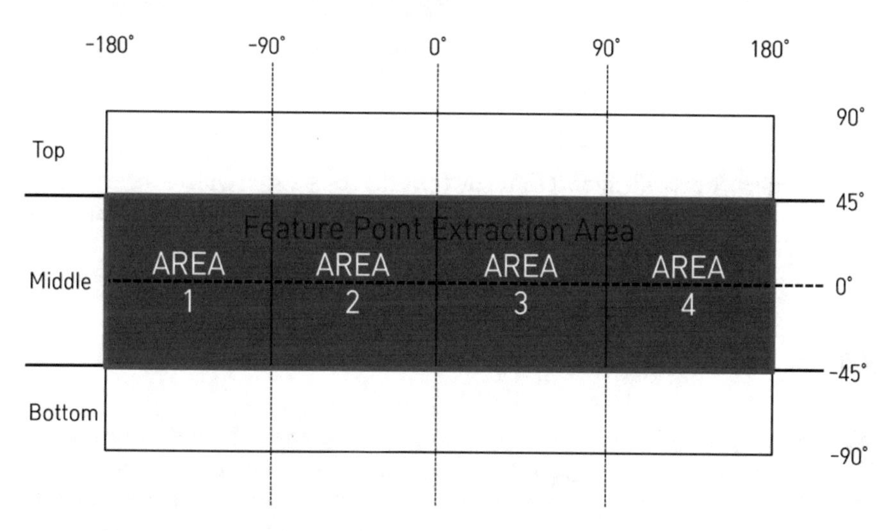

Fig. 9 ERP-based frame extraction area

Fig. 10 Area dividing and distortion correction

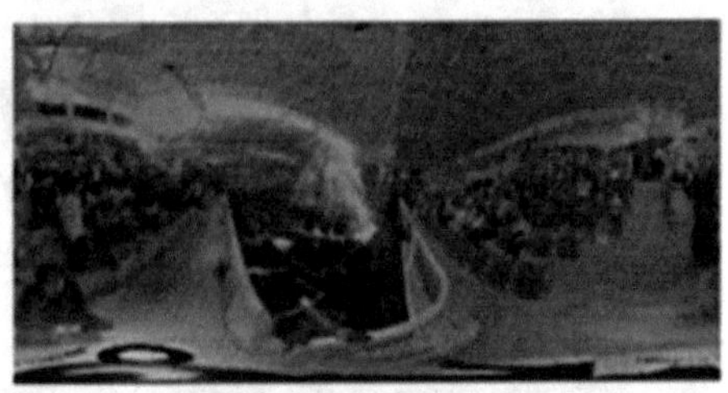
Original Frame

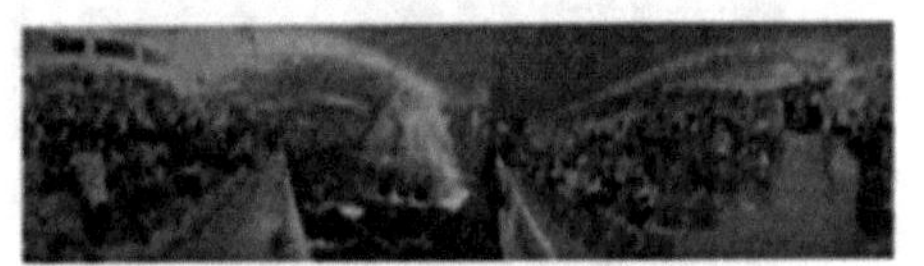
Selection of Feature Point Extraction Area

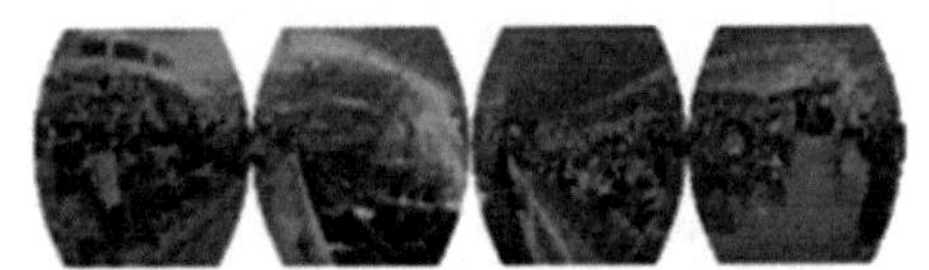
Compensation of Spherical Coordinate System

The sequence classification is different from the comparison with the query image. The speed of the query image comparison is important, but the priority is put on the accuracy for the sequence classification process (Fig. 11).

3.3 Frame Similarity Comparison

When a query image is inputted, it is necessary to compare the similarity with the image of the comparison object. When comparing the similarity, we compare the representative frame of the classified sequence with the frame extracted from the query image. A representative frame of each sequence and a region extracted from the query image are designated, and the feature points are extracted from the frame in which the distortion correction is completed and the similarity is compared.

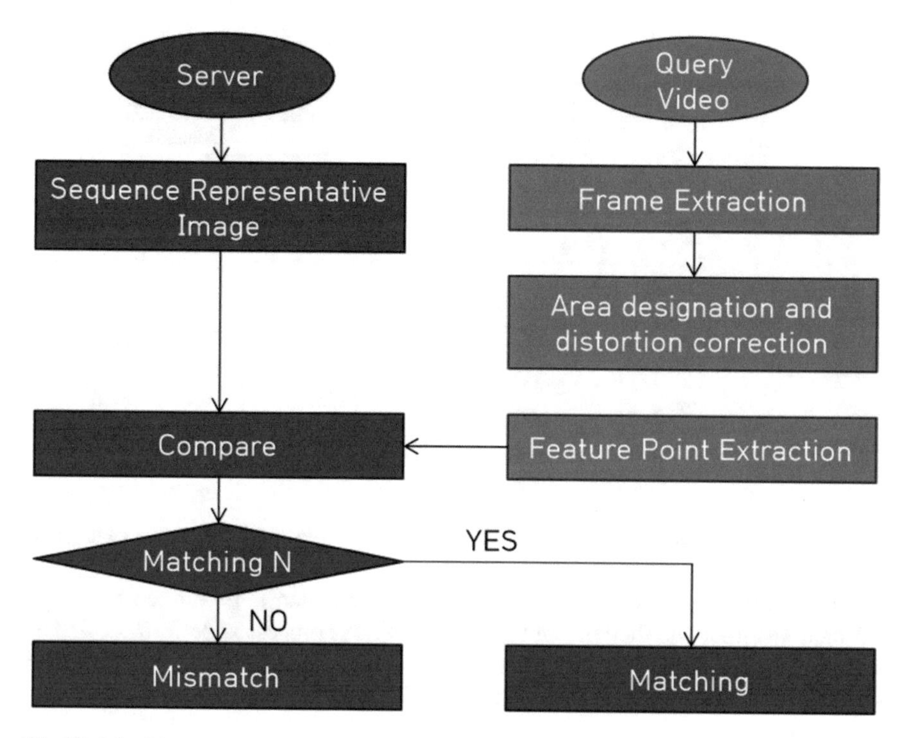

Fig. 11 Matching process

4 Experimental Results

In order to classify the sequence of VR contents, the number of matching is fixed and it is confirmed whether the desired sequence is actually classified. A sequence is classified by extracting a frame from one VR content.

From the extracted frame shown in Fig. 12, we can judge that frames from 001 to 003 are same and frames from 004 to 006 are same. We confirmed that it is actually judged through the ORB algorithm (Table 1).

From the results of the experiment, we were able to confirm that the number of matches matched at around 200 and that they were classified in the same sequence. When the number 200 were crossed, it is confirmed that there would be more number of mismatched frames as the number of matching increases since the movements

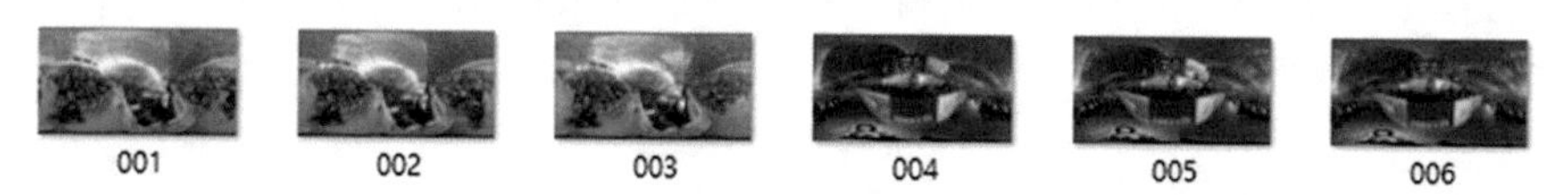

Fig. 12 Extracted frames

Table 1 Sequence classification results

Matching count	001	002	003	004	005	006
100	Matching			Matching		
150	Matching			Matching		
200	Matching			Matching		
250	Mismatch			Mismatch		
300	Mismatch			Mismatch		

Table 2 Experimental results

Experiment	(A)	(B)	(C)	(D)
①	823.7	841.1	198.2	4.1
②	1305.1	1453.7	250.4	7.2
③	701.6	684.8	150.9	3.9

of people in frames are slightly different although they had the same place and background.

In the comparison process, ① the distortion correction is performed by designating the region, and then the similarity comparison process is performed. ② The similarity comparison is performed with respect to the images which do not correct the distortion without specifying the region. ③ The performance is evaluated using the results.

When a query image is inputted, a frame of the query image is extracted and the similarity of the representative images of the classified sequences is compared.

The performance evaluation items in Table 2 are (A) the average number of feature points extracted from each extracted from the original image, (B) the average number of feature points extracted from the compared image, (C) the average number of matched feature points between the original image and the compared image, (D) the average processing time from extraction to matching.

Sequences were classified and VR images were extracted, representative frames were extracted, and ORB algorithm was used to show the number of features, the number of matches, and the time from the extraction to the matching.

The results of experiment ① and ③ showed that although the number of feature extraction between the original and the compared images was not much different, both the number of extracted frames for comparison and the number of matched count in experiment ①, which used the proposed method, were increased and that it could yield more accurate comparison for similarity. The experiment ② was performed on the entire high-definition image and may yield better accuracy for similarity comparison since it had more number of extracted frames and matched count, but resulted in more time.

5 Conclusion

In this paper, we propose a feature recognition technology for VR contents to protect the copyright of VR contents. In order to solve the performance problem of image recognition due to characteristics of 360° realistic contents, the proposed method excludes heavily distorted area such as the top and bottom parts of the image, divides the less-distorted area into 4 regions, corrects the distortion and then extracts the feature points from the corrected region. Since the extracted images are connected to successive scenes, similar frames tend to be extracted.

The frames extracted in this way are used to generate representative images of sequences that are similar to each other for speed and accuracy when comparing images in the future. As shown in the experiments, the proposed method using the representative image of the selected sequence is superior in speed for the image comparison than other methods, and it is also advantageous when the data to be stored in the server increase in the future.

For the future research, it is necessary to study the comparing method for unequal images, and the method to improve robustness and speed to guarantee the identity even after the original image is transformed.

Acknowledgements This research project was supported by Ministry of Culture, Sport and Tourism (MCST) and Korea Copyright Commission in 2019 (2018-360_DRM-9500).

References

1. Chun, H.W., Han, M.K., Jang, J.H.: Application trends in virtual reality. Electron. Telecommun. Trends (2017)
2. Yoon, J.S.: The study on the commons of copyright works. J. Inf. Law **10**, 1–44 (2016)
3. Kim, J.Y.: Design of 360 degree video and VR contents. Communication Books (2017)
4. Kijima, R., Yamaguchi, K.: VR device time - Hi-precision time management by synchronizing times between devices and host PC through USB. In: IEEE Virtual Reality (VR) (2016)
5. Ho, Y.S.: MPEG-I standard and 360 degree video content generation. J. Electr. Eng. (2017)
6. W16824, Text of ISO/IEC DIS 23090-2 Omnidirectional Media Format (OMAF)
7. Ha, W.J., Sohn, K.A.: Image classification approach for Improving CBIR system performance. In: 2016 KICS Conference (2016)
8. Song, J.S., Hur, S.J., Park, Y.W., Choi, J.H.: User positioning method based on image similarity comparison using single camera. J. KICS **40**(8), 1655–1666 (2015)
9. Yasmin, M., Mohsinm, S., Irum, I., Sharif, M.: Content based image retrieval by shape, color and relevance feedback. Life Sci. J. **10**, 593–598 (2013)
10. Everingham, M., et al.: The pascal visual object classes (voc) challenge. Int. J. Comput. Vis. **88**(2), 303–338 (2010)
11. Ke, Y., Sukthankarm, R.: PCA-SIFT: a more distinctive representation for local image descriptors. IEEE Comput. Soc. Conf. CVPR **2**, 2014 (2004)
12. Jung, H.J., Yoo, J.S.: Feature matching algorithm robust to viewpoint change. KECS **40**(12), 2363–2371 (2015)
13. Woo, N.H.: Research on the improvement plan for the protection and use activation of image copyright. Korea Sci. Art Forum **15**, 233–246 (2014)

14. Ren, S., He, K., Girshick, R., Sun, J.: Faster R-CNN: towards real-time object detection with region proposal networks. IEEE Trans. Pattern Anal. Mach. Intell. **29**(6), 1137–1149 (2017)
15. Lowe, D.G.: Distinctive image features from scale-invariant keypoints. IJCV (2004)
16. Viola, P., Jones, M.: Rapid object detection using a boosted cascade of simple features. In: Proceedings of IEEE Computer Society Conference on Computer Vision and Pattern Recognition, Kauai Hawaii (2001)
17. Rosten, E., Drummond, T.: Machine learning for high-speed corner detection. In: European Conference on Computer Vision ECCV, pp. 430–443 (2006)
18. Lowe, D.: Distinctive image features from scale-invariant keypoints. Int. J. Comput. Vis. **60**, 91–110 (2004)
19. Ke, Y., Sukthankar, R.: PCA-SIFT: a more distinctive representation for local image descriptor. In: IEEE CVPR (2004)
20. Bay, H., Tuytelaars, T., Van Gool, L.: SURF: speeded up robust features. In: ECCV (2006)
21. Bay, H., Tuytelaars, T., Van Gool, L.: SURF: speeded-up robust features. Comput. Vis. Image Underst. **110**(3), 346–359 (2008)
22. Lienhart, R., Maydt, J.: An extended set of haar-like features for rapid object detection. In: IEEE ICIP 2002, vol. 1, pp. 900 (2002)

Dimension Reduction by Word Clustering with Semantic Distance

Toshinori Deguchi and Naohiro Ishii

Abstract In information retrieval, Latent Semantic Analysis (LSA) is a method to handle large and sparse document vectors. LSA reduces the dimension of document vectors by producing a set of topics related to the documents and terms statistically. Therefore, it needs a certain number of words and takes no account of semantic relations of words. In this paper, by clustering the words using semantic distances of words, the dimension of document vectors is reduced to the number of word-clusters. Word distance is able to be calculated by using WordNet. This method is free from the amount of words and documents. For especially small documents, we use word's definition in a dictionary and calculate the similarities between documents.

1 Introduction

In information retrieval, a document is represented as a vector of index terms, which is called a document vector, and a set of documents is represented as a term-document matrix arranging document vectors in columns.

Usually, term-document matrices become large and sparse. Therefore, it is popular to use Latent Semantic Analysis (LSA) [1], Probabilistic Latent Semantic Analysis (pLSA), or Latent Dirichlet Allocation (LDA) to reduce the dimension of document vectors to find the latent topics by using statistical analysis. These methods are based on a large corpus of text, and the words are treated statistically but not semantically. Because the words are treated literally, for example, 'dog' and 'canine' are treated as totally different words, unless they co-occur in several documents.

Morave et al. [2] proposed a method to use WordNet [3] ontology. They used ℓ top level concepts in WordNet hierarchy as the topics, which means the topics

T. Deguchi (✉)
National Institute of Technology, Gifu College, Gifu, Japan
e-mail: deguchi@gifu-nct.ac.jp

N. Ishii
Department of Information Science, Aichi Institute of Technology, Toyota, Japan

R. Lee (ed.), *Big Data, Cloud Computing, and Data Science Engineering*,
Studies in Computational Intelligence 844,
https://doi.org/10.1007/978-3-030-24405-7_10

are produced semantically. For given ℓ, the concepts that is used for the topics are determined independently of the target documents.

In this paper, we propose a semantical method to reduce the dimension of term-document matrices. Instead of latent topics by statistics or ℓ top level concepts in WordNet, this method uses word-clusters generated by hierarchical clustering. For hierarchical clustering, we use a semantic distance between words, which is able to be calculated by using WordNet. The dimension of document vectors is reduced to the number of word-clusters.

When the number of documents or words are not large, statistical method is not suitable, whereas the word-clustering method uses semantic distances of words and is applicable even to that case.

For demonstration, we take especially small documents which are word's definitions in a dictionary, and show the results with vector-space, LSA, WordNet concepts, and word-clustering method.

2 Vector Space Model

In vector space model, a document is represented as a vector of terms (or words). For the values of a document vectors, term frequency (tf) or term frequency-inverse document frequency (tf-idf) is used to show the importance of the word in the document in the collection.

Although there are several definitions for tf and idf, in this paper, we use

$$\text{tf-idf}(t, d) = \text{tf}(t, d) \times \text{idf}(t), \tag{1}$$

$$\text{tf}(t, d) = n_{t,d}, \tag{2}$$

$$\text{idf}(t) = \log \frac{|D|}{df_t}, \tag{3}$$

where $n_{t,d}$ is the raw count of the word w_t in the document doc_d, D is the set of document, and df_t is the number of documents where the word w_t appears.

A term-document matrix (or document matrix) is a matrix that is constructed by lining up document vectors as column vectors (Fig. 1). Therefor, for t terms in d document, the matrix has the dimension of $t \times d$.

Fig. 1 Term-document matrix X

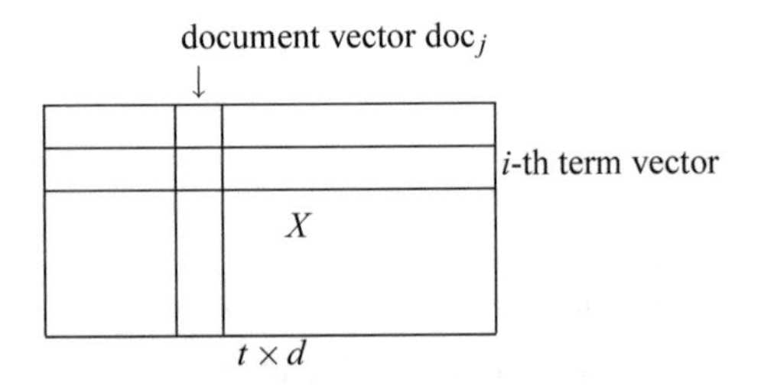

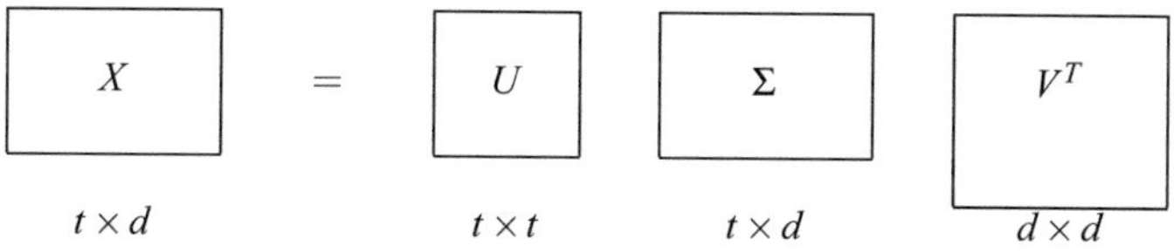

Fig. 2 Singular value decomposition of the term-document matrix X

Fig. 3 Singular value decomposition of the term-document matrix X

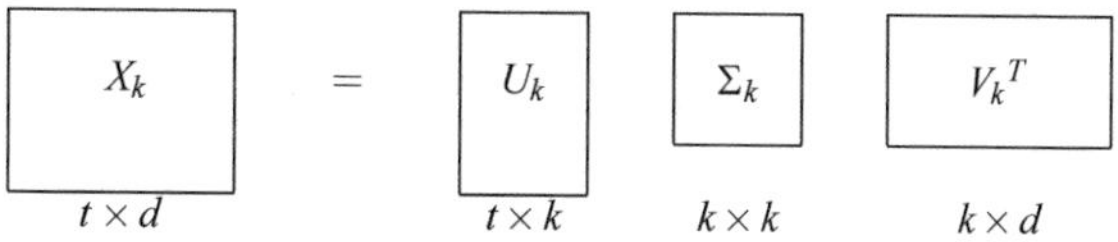

The document similarities are calculated as follows:

$$\text{similarity}(\text{doc}_i, \text{doc}_j) = \frac{\text{doc}_i \cdot \text{doc}_j}{\|\text{doc}_i\| \ \|\text{doc}_j\|} \tag{4}$$

3 LSA

A term-document matrix is a large and sparse matrix. LSA [1] is a method to reduce the dimension of document matrix and find topics in the documents, calculating topic-document matrix with singular-value decomposition (SVD).

With SVD, term-document matrix X is factorized as follows:

$$X = U \Sigma V^T, \tag{5}$$

where U is $t \times t$ matrix, Σ is $t \times d$ rectangular diagonal matrix—suppose the values are listed in descending order, and V is $d \times d$ matrix as in Fig. 2.

Let Σ_k be composed of first k columns and first k rows of Σ. Let U_k and V_k be composed of first k columns of U and V respectively. Then, the term-document matrix X is approximated by the matrix:

$$X_k = U_k \Sigma_k {V_k}^T, \tag{6}$$

as in Fig. 3.

To reduce the dimension of X, the topic-document matrix Y is calculated as:

$$Y = {U_k}^T X = \Sigma_k {V_k}^T \tag{7}$$

Table 1 An example of hierarchies in WordNet

Depth	Concept	
1	entity	
2	physical_entity	
3	physical_object	
4	whole	
5	living_thing	
6	being	
7	animate_being	
8	chordate	
9	vertebrate	
10	mammalian	
11	placental	
12	aquatic_mammal	hoofed_mammal
13	blower	odd-toed_ungulate
14	whale	equine
15		horse

4 WordNet Concepts

Because LSA is statistical, it is on the assumption that there are a lot of words and documents. When there are not so many words and/or documents, a statistical method is not suitable.

Instead of LSA topics, a method using WordNet hypernyms in n top levels is proposed [2]. WordNet [3] is a lexical database. In WordNet, nouns and verbs are arranged in concept (synset in WordNet) hierarchies by hypernyms. For example, a part of hierarchies including the concepts of the word "whale" and the word "horse" is shown in Table 1.

Therefore, the concepts within n top levels can represent topics. Moravec et al. wrote in [2] as follows:

> We can use all hypernyms of given term from ℓ top levels, applying a fraction of term weight dependent on its level in WordNet to hypernym weight in concept-by-document matrix.

The term weight in a concept was inversely proportional to a logarithm of concept level.

Since there's no description about multi-synsets or multi-paths, We treat these as follows. For multi-synsets for one word, we use every synsets as concepts to generate the concept-word matrix $A = (a_{i,j})$. For multi-paths from one synset, we take the one which gives the maximum coefficient.

Let s be a synset.

$$\text{term-weight}(s) = \frac{1}{\log(\text{depth}(s)) + 1}, \tag{8}$$

where $\text{depth}(c) = \text{len}(c, \textit{root})$—a global root *root* is imposed above the hierarchies to ensure that there is a path between every two synsets [4], and $\text{len}(c_i, c_j)$ is the length of shortest path in WordNet from synset c_i to synset c_j measured in edges. Note that we add $+1$ to the denominator for the case of $\text{depth}(s) = 1$.

Let S_j be a set of synsets for word w_j. For a synset t_i in ℓ top levels and a synset s, we define the fraction of synset weight to hypernym weight as:

$$f(t_i, s) = \begin{cases} 1, & \text{for } s \in \bigcup_{k=1}^{\ell} T_k, \\ \dfrac{\text{term-weight}(s)}{\text{term-weight}(t_i)}, & \text{for } s \in \text{sbo}(t_i),\ t_i \in T_\ell, \\ 0, & \text{for otherwise}, \end{cases} \tag{9}$$

where T_ℓ is a set of synsets at ℓ top levels, which means $\text{depth}(t) = \ell$ for all $t \in T_\ell$ and $\text{sbo}(t)$ is a set of synsets which are subordinate of t. For the fraction of term weight to hypernym weight, we take the maximum f of all the synset $s \in S_j$.

$$a_{i,j} = \max_{s \in S_j} f(t_i, s) \tag{10}$$

Then, we can reduce the word-document matrix X to the concept-document matrix Y as:

$$Y = AX \tag{11}$$

The disadvantage of this method is not to chose a number of topics or a range of concepts arbitrarily. Then, we use word-clustering to construct topics and reduce the dimension.

5 Word Clustering with Semantic Distance

By dividing the words in the documents to clusters, the term-document matrix is mapped to cluster-document matrix. For clustering, we use hierarchical clustering with semantic distance of words. Semantic distance of words is able to be calculated by using WordNet.

5.1 Distance by WordNet

There are several definitions to measure semantic relatedness using WordNet. Among them, we adopt Wu and Palmer's method [5] because the similarity value of two words is from 0 to 1 and the distance is calculated as subtracting the similarity from 1.

According to Budanitsky and Hirst [4], the relatedness $\text{rel}(w_1, w_2)$ between two words w_1 and w_2 can be calculated as

$$\text{rel}(w_1, w_2) = \max_{c_1 \in s(w_1), c_2 \in s(w_2)} [\text{rel}(c_1, c_2)], \tag{12}$$

where $s(w_i)$ is the set of concepts in WordNet that are senses of word w_i. We consider the relatedness as the similarity of two words as:

$$\text{sim}(w_1, w_2) = \max_{c_1 \in s(w_1), c_2 \in s(w_2)} [\text{sim}(c_1, c_2)] \tag{13}$$

In Wu and Palmer's method, the similarity of two concepts is defined as follows:

$$\text{sim}(c_i, c_j) = \frac{2 \text{ depth}(\text{lso}(c_i, c_j))}{\text{len}(c_i, \text{lso}(c_i, c_j)) + \text{len}(c_j, \text{lso}(c_i, c_j)) + 2 \text{ depth}(\text{lso}(c_i, c_j))} \tag{14}$$

where $\text{lso}(c_1, c_2)$ is the lowest super-ordinate (or most specific common subsumer) of c_1 and c_2.

For example, from Table 1, letting c_1 = whale and c_2 = horse,

$$\text{lso}(\text{whale}, \text{horse}) = \text{placental}, \tag{15}$$

$$\text{depth}(\text{placental}) = 11, \tag{16}$$

$$\text{len}(\text{whale}, \text{placental}) = 3, \tag{17}$$

$$\text{len}(\text{horse}, \text{placental}) = 4, \tag{18}$$

$$\text{sim}(\text{whale}, \text{horse}) = \frac{2 \times 11}{3 + 4 + 2 \times 11} \simeq 0.759. \tag{19}$$

In computation of $\text{sim}(c_i, c_j)$, there are cases that more than one synset for $\text{lso}(c_i, c_j)$ exists, For example, if you take water and woodland, the hierarchy of these concepts is obtained as shown in Table 2. Because depth(physical_entity) = depth(abstract_entity) = 2, lso(water, woodland) can be both physical_entity and abstract_entity.

If you use physical_entity for lso(water, woodland), you will get

$$\text{sim}(\text{water}, \text{woodland}) = \frac{2 \times 2}{4 + 3 + 2 \times 2} \simeq 0.364. \tag{20}$$

Otherwise,

$$\text{sim}(\text{water}, \text{woodland}) = \frac{2 \times 2}{6 + 4 + 2 \times 2} \simeq 0.286. \tag{21}$$

Table 2 Hierarchy on water and woodland

Depth	Concept	
1	entity	
2	physical_entity	
3	physical_object	
4	whole	
5	living_thing	
6	being	
7	animate_being	
8	chordate	
9	vertebrate	
10	mammalian	
11	placental	
12	aquatic_mammal	hoofed_mammal
13	blower	odd-toed_ungulate
14	whale	equine
15		horse

In these cases, we take the maximum as follows:

$$\mathrm{sim}(c_i, c_j) = \max_{\mathrm{lso}(c_i,c_j)} \left[\frac{2\ \mathrm{depth}(\ \mathrm{lso}(c_i, c_j))}{\mathrm{len}(c_i,\ \mathrm{lso}(c_i, c_j)) + \mathrm{len}(c_j,\ \mathrm{lso}(c_i, c_j)) + 2\ \mathrm{depth}(\ \mathrm{lso}(c_i, c_j))} \right]. \tag{22}$$

Equation (22) is determined uniquely even when there are multi-paths from c_i or c_j to $\mathrm{lso}(c_i, c_j)$, since we defined $\mathrm{len}(c_i, c_j)$ as the length of the shortest path from c_i to c_j.

From above, we also get the definition of the distance of two words as follows:

$$d(w_i, w_j) = 1 - \mathrm{sim}(w_i, w_j) \tag{23}$$

5.2 *Word Clustering*

First, distance matrix $D = (d_{ij})$ for all words in term-document matrix is calculated as:

$$d_{ij} = d(w_i, w_j) \tag{24}$$

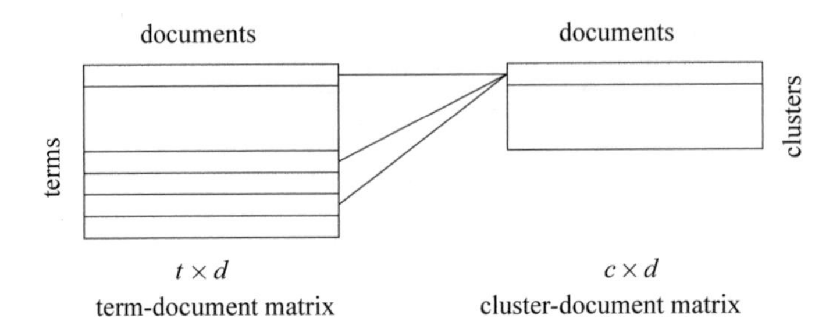

Fig. 4 Dimension reduction by word clustering

Then, using the matrix D, hierarchical clustering divides the words into clusters by cutting the dendrogram to form c flat clusters. We use Ward's method as the linkage criteria. For the following calculations, the cluster membership matrix $M = (m_{ij})$ is formed as:

$$m_{ij} = \begin{cases} 1, \text{ for } w_j \in C_i, \\ 0, \text{ for } w_j \notin C_i \end{cases} \tag{25}$$

5.3 *Dimension Reduction*

Using the clusters as topics in LSA, we can reduce the dimension of the term-document matrix and turn it into the cluster-document matrix, joining the words in the same cluster as in Fig. 4.

In order to convert the matrix, we need the mapping functions from the words to the clusters. In the same way as in LSA, let us assume that each function is represented by a linear combination of the words and that the mapping is represented by a cluster-word relation matrix $R = (r_{i,j})$. Then, we can calculate the cluster-document matrix Y as follows:

$$Y = RX \tag{26}$$

The details of R are described in the next section.

6 Experimental Results

6.1 *Conditions of Experiments*

For the small documents, we use the first definition of the 6 words *dog*, *cat*, *horse*, *car*, *bus*, and *ship* on online "Cambridge Dictionary" [6]. Each definition sentence is a document. We describe the document defining *dog* as 'dog', and so on.

Table 3 Execution environment

Language	Python 3.7.1
Stemming	TreeTagger [7]
POS tagging	TreeTagger
LSA	TruncatedSVD (in scikit-learn 0.19.2)
WordNet	Japanese WordNet 1.1 [8]
Hierarchical clustering	Hierarchy (in scipy 1.1.0)
Linkage method	Ward's method

Note Japanese WordNet 1.1 includes Princeton WordNet 3.0.

We use the environment in Table 3. The distances between words are calculated by our tool according to (13), (22), and (23).

WordNet has hypernym hierarchies only for nouns and verbs. Therefore, the input texts are divided into stem words which are POS-tagged, then only nouns and verbs are used to generate the document matrix. For stemming and POS-tagging, we use TreeTagger because it performs the best for tagging nouns and verbs [9].

This matrix is used in all methods including LSA.

6.2 *Cluster-Word Relation Matrix*

To calculate the mapping function from the words to the clusters, we define the grade of membership of the word to the cluster. For word w_j, the grade $g(C_i, w_j)$ of membership to cluster C_i is considered to have a relation to the distances between w_j and the words in cluster C_i. Although the inverse proportion to the distance is popular, we will have the multi-synsets problem in which two different word w_j and w_k could be $d(w_j, w_k) = 0$ and the grade becomes infinity. To modify the inverse proportion is one plan, but we chose to use similarity of words.

We define the grade $g(C_i, w_j)$ as the mean similarity of the words as follows:

$$g(C_i, w_j) = \frac{1}{|C_i|} \sum_{w \in C_i} \mathrm{sim}(w, w_j) \tag{27}$$

Then, we simply use this grade for the cluster-word relation matrix.

$$r_{i,j} = g(C_i, w_j) \tag{28}$$

For calculating the matrix, we need word similarity matrix $S = (s_{ij})$,

$$s_{ij} = 1 - d(w_i, w_j). \tag{29}$$

Table 4 Similarities by vector space

	'dog'	'cat'	'horse'	'car'	'bus'
'cat'	0.229				
'horse'	0.176	0.055			
'car'	0.012	0.000	0.040		
'bus'	0.015	0.000	0.051	0.049	
'ship'	0.000	0.000	0.000	0.000	0.000

Using S and membership matrix M, we can calculate the grades without normalization as MS.

For normalization, we prepare normalizing matrix $N = (n_{ij})$ as:

$$n_{ij} = \frac{1}{|C_j|} \tag{30}$$

Then, we can calculate cluster-word relation matrix as:

$$R = (MS) \circ N, \tag{31}$$

where $\circ$ means Hadamard product.

6.3 *Similarity of Documents*

In LSA, WordNet concepts, and word-clustering method, we get Y reducing the dimension of the document matrix as (7), (11), and (26) respectively.

As same as (4) in the vector space model, we use the cosine similarity. For the calculation of the document similarity, let $\mathbf{d}_j = (y_{1j}, \ldots, y_{dj})^T$ for $Y = (y_{ij})$. Then, the document similarity between document doc_i and document doc_j is calculated as follows:

$$\mathrm{similarity}(i, j) = \frac{\mathbf{d}_i \cdot \mathbf{d}_j}{\|\mathbf{d}_i\| \ \|\mathbf{d}_j\|} \tag{32}$$

6.4 *Result of Each Method*

Table 4 shows the document similarities with raw vector space model calculated by (4). Almost all the documents have no similarity each other. From all the documents, 30 terms of nouns and verbs are extracted, and only 7 terms appear in more than one document as shown in Table 5. Thus, almost all vectors have different directions.

Table 5 Words appeared in more the one document

Word	Documents
Animal	‘dog’, ‘cat’, ‘horse’
Leg	‘dog’, ‘cat’, ‘horse’
People	‘dog’, ‘horse’, ‘car’, ‘bus’
Pet	‘dog’, ‘cat’
Thing	‘dog’, ‘horse’
Vehicle	‘horse’, ‘car’, ‘bus’
Keep	‘dog’, ‘cat’

Table 6 Similarities by LSA (5 topics)

	‘dog’	‘cat’	‘horse’	‘car’	‘bus’
‘cat’	0.722				
‘horse’	0.693	0.011			
‘car’	0.020	0.000	0.041		
‘bus’	−0.042	0.009	0.063	0.049	
‘ship’	−0.000	−0.000	0.000	0.000	−0.000

Table 7 Similarities by WordNet concepts at 2 top level

	‘dog’	‘cat’	‘horse’	‘car’	‘bus’
‘cat’	0.579				
‘horse’	0.126	0.138			
‘car’	0.388	0.550	0.170		
‘bus’	0.112	0.129	0.173	0.182	
‘ship’	0.223	0.320	0.167	0.579	0.104

In the case that only 7 terms appear in different documents, using LSA is questionable. However, we used LSA for a comparison.

The result using LSA is shown in Table 6. The number of topics was decided to 5, as the cumulative contribution ratio was over 80%. Only similarity(‘dog’, ‘cat’) and similarity(‘dog’, ‘horse’) are high, because there are 4 common terms in ‘dog’ and ‘cat’, and 4 terms in ‘dog’ and ‘horse’. As described above, this is not the case to use statistical analyses.

The result using WordNet concept is shown in Table 7. As seen in the table, the values are too small. Inspecting the 2 top level concepts used in the result, it turned out that only 3 concepts for noun is used, while there are 138 concepts for verb. Therefore, we consider that top concepts does not represent topics in this case.

Table 8 Similarities by WordNet concepts of nouns at 4 top level

	'dog'	'cat'	'horse'	'car'	'bus'
'cat'	0.695				
'horse'	0.940	0.606			
'car'	0.639	0.670	0.658		
'bus'	0.452	0.269	0.442	0.388	
'ship'	0.275	0.427	0.325	0.390	0.186

For another trial, we used the concepts only for nouns at 4 top levels. The results is in Table 8. This is, we consider, a fairly good, because it shows high similarities between animals.

However, inspecting the words in each concept, a problem of multi-synsets was found. Table 9 shows the concepts and the words in each concept. Some words belong to several concepts and there are 22 concepts that includes one word. From these reasons, the number of concepts become 36 though the number of nouns was 19. This means that the dimension is not reduced well.

The result using word-clustering is shown in Table 10. The number of cluster is decided with several pre-experiments checking cluster members. All the terms are divided into 7 clusters as in Table 11.

From the result, it seems that similarities are too high. As we adopted (31) as a cluster-term relation matrix, a coefficient is set for all terms including the terms that do not belong to the cluster.

In the next section, we will arrange the relation matrix differently.

6.5 Variation of Cluster-Word Relation Matrix

In Table 10, the similarities are fairly large, because we used all the similarities between a cluster and term.

In this section, terms used in producing a coefficient are restricted in the cluster as shown in (33).

$$r'_{i,j} = \begin{cases} g(C_i, w_j), & \text{for } w_j \in C_i, \\ 0, & \text{for } w_j \notin C_i. \end{cases} \tag{33}$$

To generate the cluster-term relation matrix, we can use the following equation.

$$R' = ((MS) \circ M) \circ N \tag{34}$$

The result by using R' is shown in Table 12.

Or we might be able to use just $R = M$. The result is shown in Table 13.

Table 9 Words in each concept by WordNet concepts

Concept	Words in concept
Whole	Animal, boat, claw, engine, fur, leg, mouse, number, pet, road, seat, tail, thing, vehicle, water, wheel
Piece	Claw, leg, seat, tail
Engine	Engine
Substance	Fur, water
Component	Fur, place, water
Form	Leg, tail
Solid	Leg
Property	Leg, number, place
Event	Leg, number, pet, place, road, thing
State	Mouse, pet, place, thing
Individual	Mouse, pet, tail
definite_quantity	Number
Signaling	number
social_group	Number, people
Indication	Number
Collection	Number
People	People
Citizenry	People
Masses	People
Location	Place, seat
Knowledge	Place, seat, thing
auditory_communication	Place
written_communication	place
body_of_water	Sea, water
indefinite_quantity	Sea
Activity	Sea
Portion	Seat
Point	Tail
Content	Thing
Something	Thing
Stuff	Thing
Thing	Thing
Substance	Vehicle, water
Vehicle	Vehicle
Fluid	Water
Quality	Wheel

Table 10 Similarities by word-clustering with WordNet (7 clusters)

	'dog'	'cat'	'horse'	'car'	'bus'
'cat'	0.904				
'horse'	0.918	0.678			
'car'	0.789	0.963	0.519		
'bus'	0.960	0.778	0.979	0.657	
'ship'	0.884	0.968	0.695	0.961	0.809

Table 11 Cluster members with WordNet (7 clusters)

Cluster no.	Terms
1	Fur, number, place, road, sea, seat, thing, water
2	People
3	Animal, mouse, pet, tail
4	Boat, claw, engine, leg, vehicle, wheel
5	Be, carry, guard, keep, use
6	Drive, ride, travel
7	Catch, hunt, pull

Table 12 Similarities by word-clustering with R' (7 clusters)

	'dog'	'cat'	'horse'	'car'	'bus'
'cat'	0.797				
'horse'	0.811	0.523			
'car'	0.344	0.402	0.403		
'bus'	0.491	0.264	0.829	0.553	
'ship'	0.301	0.323	0.550	0.858	0.817

Table 13 Similarities by word-clustering with M (7 clusters)

	'dog'	'cat'	'horse'	'car'	'bus'
'cat'	0.740				
'horse'	0.876	0.519			
'car'	0.322	0.451	0.363		
'bus'	0.596	0.329	0.827	0.578	
'ship'	0.289	0.379	0.457	0.892	0.773

These results are quite good, in our point of view, because the similarities between the animals are high and the ones between the transporters including 'horse' except the one between 'horse' and 'car'.

7 Conclusion

Using word-clustering by semantic distance derived from WordNet, we transformed term-document matrix to cluster-document matrix. This transformation is a dimension reduction of document vectors.

We demonstrated this method to calculate the similarities of the small documents. Unfortunately, we have no quantitative evaluation, but each result of the experiments suggests that this method is advantageous for a small set of documents. To show that this method can be applicable to a large set, we need more examinations.

We can also use Word2Vec [10, 11] to calculate the distance of words. Although Word2Vec needs a large amount of documents to train a model, the use of pre-trained model can be separated from training the model. For example, we can use pre-trained model of GloVe [12]. Use of Word2Vec and its verification are issues in the future.

References

1. Deerwester, S., Dumais, S.T., Furnas, G.W., Landauer, T.K., Harshman, R.: Indexing by latent semantic analysis. J. Am. Soc. Informat. Sci. **41**(6), 391–407 (1990)
2. Moravec, P., Kolovrat, M., Snášel, V.: LSI vs. Wordnet Ontology in Dimension Reduction for Information Retrieval. Dateso, 18–26 (2004)
3. Miller, G.: WordNet: An Electronic Lexical Database. MIT Press (1998)
4. Budanitsky, A., Hirst, G.: Evaluating WordNet-based measures of lexical semantic relatedness. Computat. Linguist. **32**(1), 13–47. MIT Press (2006)
5. Wu, Z., Palmer, M.: Verbs semantics and lexical selection. In: Proceedings of the 32nd Annual Meeting on Association for Computational Linguistics, pp. 113–138 (1994)
6. https://dictionary.cambridge.org/
7. Schmid, H.: Probabilistic part-of-speech tagging using decision trees. New Methods in Language Processing, 154–164 (2013). http://www.cis.uni-muenchen.de/%7Eschmid/tools/TreeTagger/
8. Bond, F., Baldwin, T., Fothergill, R., Uchimoto, K.: Japanese SemCor: a sense-tagged corpus of Japanese. In: Proceedings of the 6th Global WordNet Conference (GWC 2012), pp. 56–63 (2012). http://compling.hss.ntu.edu.sg/wnja/index.en.html
9. Tian, Y., Lo, D.: A comparative study on the effectiveness of part-of-speech tagging techniques on bug reports. In: 2015 IEEE 22nd International Conference on Software Analysis, Evolution and Reengineering (SANER), pp. 570–574 (2015)
10. Mikolov, T., Chen, K., Corrado, G., Dean, J.: Efficient Estimation of Word Representations in Vector Space. arXiv:1301.3781 (2013)
11. Mikolov, T., Sutskever, I., Chen, K., Corrado, G., Dean, J.: Distributed representations of words and phrases and their compositionality. Advanc. Neural Informat. Process. Syst. **26**, 3111–3119 (2013)
12. Pennington, J., Socher, R., Manning, C.D.: GloVe: global vectors for word representation. Empirical Methods in Natural Language Processing (EMNLP), pp. 1532–1543 (2014). https://nlp.stanford.edu/projects/glove/

Word-Emotion Lexicon for Myanmar Language

Thiri Marlar Swe and Phyu Hninn Myint

Abstract In recent years, social media emerges and becomes mostly use all over the world. Social media users express their emotions by posting status with friends. Analysis of emotions of people becomes popular to apply in many application areas. So many researchers propose emotion detection systems by using Lexicon based approach. Researchers create emotion lexicons in their own languages to apply in emotional system. To detect Myanmar social media users' emotions, lexicon does not available thus; a new word-emotion lexicon especially based on Myanmar language is needed to create. This paper describes the creation of Myanmar word-emotion lexicon, M-Lexicon that contains six basic emotions: happiness, sadness, fear, anger, surprise, and disgust. Facebook status written in Myanmar words are collected and segmented. Words in M-Lexicon is finally got by applying stop-words removal process. Finally, Matrices, Term-Frequency Inversed Document Frequency (TF-IDF), and unity-based normalization is used in lexicon creation. Experiment shows that the M-lexicon creation contains over 70% of correctly associated with six basic emotions.

Keywords Word-emotion lexicon · M-Lexicon · TF-IDF · Unity-based normalization

1 Introduction

Social media becomes most popular communication median among users. People spend a lot of time on social media by positing texts, images, and videos to express their emotions, to get up-to-date news and other interested information. The opinions and emotions of social media users become important in many situations such as

T. M. Swe · P. H. Myint (✉)
University of Computer Studies, Mandalay, Myanmar
e-mail: phyuhninmyint@ucsm.edu.mm

T. M. Swe
e-mail: thirimarlarswe@gmail.com

R. Lee (ed.), *Big Data, Cloud Computing, and Data Science Engineering*,
Studies in Computational Intelligence 844,
https://doi.org/10.1007/978-3-030-24405-7_11

elections, medical treatment, etc. In order to identify user's emotional situation from the social media, many researchers propose emotion detection system from text in so many techniques. Some researchers propose and create emotion lexicon to detect emotional status from the text in their own languages such as English, France, Arabic, Chinese, Korean, etc.

For Myanmar, there is no emotion lexicon which to be applied in detecting emotion words from Myanmar social media users' status. Language translation tools and existing different language emotion lexicon can be used for Myanmar language, but it is not sufficient for practical system. In addition, an emotion lexicon can be manually built, but it is time-consuming. Therefore, a new word-emotion lexicon for Myanmar language, namely M-Lexicon, is programmatically created. This lexicon contains six basic emotions such as happiness (joy), sadness, fear, anger, surprise, and disgust as defined by Ekman [1, 2, 3].

In this paper, the creation process of M-Lexicon is presented. This lexicon is created with words from a social media, Facebook. Although there has different Facebook status such as text status, video and image status, two kinds of status: Myanmar text only and status with feelings are applied in the lexicon creation process.

Before creating M-Lexicon, the Facebook status text is segmented in order to get words from it. In the segmentation process, there are two phases: syllable segmentation and syllable merging. The segmented status may contain stop words, which are unnecessary words in the creation of M-Lexicon. Thus, there is also needed to remove stop words from the segmented status. The emotion lexicon is then created by using the matrices and preprocessed words or terms. Term-Frequency Inversed Document Frequency (TF-IDF) scheme is applied in matrix computational process and finally unity-based normalization is performed on the matrix values to adjust it in range [0, 1].

The rest of this paper is organized as follows. In Sect. 2, the related works are presented. Section 3 describes Myanmar language processing which is needed to perform before M-Lexicon creation. This paper explains how to create an M-Lexicon in Sect. 4. Section 5 discusses experimental results followed by conclusion in Sect. 6.

2 Related Works

Emotion lexicons are constructed based on different methodologies and languages. Some researchers proposed an automatic way while other ones are generating it in manual, translating the existing lexicon to their own languages.

Bandhakavi et al. [4] proposed a set of methods to automatically generate a word-emotion lexicon from the emotional tweets corpus. They used term frequencies for learning lexicons and more sophisticated methods that are based on conceptual models on how tweets are generated. Their methods obtain significantly better scores. However, their methods can only work on short text representation and cannot get better results on other domains such as blogs, discussion forums wherein the text size is larger than tweets.

Although many researchers used automatic methods to assist the lexicon building procedure. Their works commonly rank words by calculating similarities values with a set of seed words. The most similar words are added to the final lexicon or used as additional seed words. Xu et al. [5] adopted a graph-based algorithm to build Chinese emotion lexicons and incorporate multiple resources to improve the quality of lexicons and save human effort. They used graph-based algorithm to rank words automatically according to a few seed emotion words. Although, they built lexicon for five emotions (happiness, anger, sadness, fear, surprise), the size of an emotion lexicon is small and the quality is not good.

Do et al. [6] presented a method to build fine-grained emotion lexicons for Korean language automatically from the annotated Twitter corpus without using other existing lexical resources. Their proposed method, weighted tweet frequency (weighted TwF), is aggregated same emotion label tweets in one document, and produce six documents as a result. Then they calculate the weighted TwF values for each term that appears in each six documents. The higher TwF values, the stronger the corresponding emotion. Their classification performance can improve by adding several fine-grained features as they suggested.

Mohammad et al. [7] utilized crowd sourcing to build and generate a large, high-quality, word-emotion and word-polarity association lexicon quickly and inexpensively. Their lexicon, NRC word-emotion association lexicon also known as EmoLex lexicons, assigns 14182 words into eight emotion categories [8], and two sentiments (positive and negative). Their lexicon mostly affects the annotations for English words. Each word in the lexicon is assigned to one or more emotion categories that are annotated manually by Mechanical Turk users.

3 Myanmar Language

This section describes the introduction to Myanmar language and explains the word-segmentation and stop-words removal process. These processes are needed to be performed before creating an M-Lexicon; more detail process of such lexicon creation is described in next section.

3.1 Introduction to Myanmar Language

Myanmar (Burmese) language, an official language in Myanmar, is the Sino-Tibetan language spoken by 30 million people in Myanmar. According to history, Myanmar script draws its source from Brahmi script which flourished in India from about 500 B.C. to over 300 AD [9, 10].

In Myanmar script, the sentences are written from left to right and clearly delimitated by a sentence boundary marker. There is no regular inter-word spacing, although inter-phrase spacing may sometimes to be used [11]. Segmentation of sentence to words become challenging tasks. Myanmar NLP team and other researchers work hard on the segmentation process of Myanmar Language.

As described in Sect. 1, the lexicon creation process mainly focus on the text resources from Facebook Users' status and the collected text status is needed to be segmented. Segmentation on Myanmar text will be explained in the following sub-section.

3.2 *Word Segmentation*

Word segmentation is used to determine the boundaries of words for those languages without word separators in orthography is a basic task in natural language processing. In linguistics, a word is a basic unit of language separated by a space. To process text, words have to be determined. For instance, emotion detector requires text source to be indexed by words. Therefore, word segmentation becomes an essential pre-requirement in applications where data and information are to be computationally processed in their language [11].

Myanmar writing does not use white spaces between words or between syllables. Due to this, word segmentation process of other language is not straightforward for Myanmar language. In this lexicon creation, Myanmar text in Facebook users' status is segmented before processing. The segmentation has two steps: syllable segmentation and syllable merging.

A syllable is a basic sound unit or a sound. A word can be made up of one or more syllables. Every syllable boundary can be a potential word boundary. A Myanmar syllable consists of one initial consonant, zero or more medial, zero or more vowels and optional dependent various signs [12]. Syllable segmentation can be performed by using a rule-based heuristic approach and the segmented syllables can be merged into words with the help of a dictionary-based statistical based approach. Table 1 describes an example of syllable segmentation and syllable merging.

Table 1 Syllable segmentation and syllable merging

Input text:	မိုးရွာနေတယ် (It is raining)
Segmented syllables:	မိုး_ရွာ_နေ_တယ်
Merged syllables:	မိုးရွာ_နေ_တယ်

3.3 Stop Words

After segmenting the sentence, each word may also be stop word. Stop word is a word that commonly appears to as suffixes to other words. Stop words are usually pronouns, prepositions, conjunctions and particles. Those words carry no emotion information and are not necessary in lexicon creation. Therefore, it needed to perform the stop words removal process first. They form closed classes and hence can be listed. Hopple [13] also noticed that particles ending phrases could be removed to recognize words in a sentence. In this system, the stop words are collected by analyzing the Facebook status. Table 2 lists some stop words of Myanmar language.

4 Word-Emotion Lexicon

Myanmar word-emotion lexicon, M-Lexicon, is created using the words from Facebook status. Word-emotion lexicon is formed as matrix. The columns express words and six basic emotions: happiness, sadness, fear, anger, surprise, and disgust. The rows are the identified emotion value for each word. The value 1 indicates that the word is associated with the emotion; otherwise, it is not. In order to get the word and identify the emotion type, the process will be systematic as mentioned in the following sections.

4.1 Data Collection

To build the M-Lexicon, the data is collected manually from a social media, Facebook. Text status is only used in lexicon creation. The six Facebook reactions: like, love, ha ha, wow, sad, angry are also taken. The 'Fear' and 'Disgust' reaction values of a status are separately collected from the users. In this paper, the lexicon is initially built using 485 Facebook status.

4.2 Data Preprocessing

Data preprocessing on collected data is carried out. Firstly, the system finds and removes status, which only contains emoticon and/or sticker. After filtering the status, Myanmar text in status needs to be segmented as described in Sect. 3. In order to do so, Word-breaker [14] is used. Example 1 shows the segmented result of a status.

Table 2 Some Myanmar stop words

Myanmar Stop Words	In English
အပေါ်၊ အနက်၊ အမြဲတမ်း၊ အတွင်းတွင်၊ မကြာမီ၊ မတိုင်မီ၊ ဒါ့အပြင်၊ အောက်မှာ၊ အထဲမှာ၊ ဘယ်တော့မျှ၊ မကြာခဏ၊ စဉ်တွင်၊ နှင့်အတူ၊ နှင့်၊ အတူတကွ	above, among, always, between, before, beside, down, inside, never, often, quite, while, with
ကျွန်တော်၊ ကျွန်မ၊ ငါ၊ ကျုပ်၊ ကျွန်ုပ်၊ ကျနော်၊ ကျမ၊ သူ၊ သူမ၊ ထိုဟာ၊ ထိုအရာ၊ ထို၊ ၎င်း၊ ကျွန်တော်တို့၊ ကျွန်မတို့၊ ငါတို့၊ ကျုပ်တို့၊ ကျွန်ုပ်တို့၊ ကျနော်တို့၊ ကျမတို့၊ သင်၊ သင်တို့၊ နင်၊ နင်တို့၊ မင်း၊ မင်းတို့၊ သူတို့၊ အဲလူတွေ၊ အဲမိန်းမတွေ၊	I, you , he, she, it, we, you, they
ကျွန်တော်အား၊ ကျွန်တော်ကို၊ ကျွန်မကို၊ ငါ့ကို၊ ကျုပ်ကို၊ ကျွန်ုပ်ကို၊ သူ့ကို၊ သူမကို၊ ထိုအရာကို၊ သင့်ကို၊ သင်တို့ကို၊ မင်းကို၊ မင်းတို့ကို၊ နင့်ကို၊ နင်တို့ကို၊ ငါတို့ကို၊ ကျုပ်တို့ကို၊ ကျွန်ုပ်တို့ကို၊ ငါတို့တွေကို၊ မင်းတို့တွေကို၊ သားကို၊ သမီးကို	me, you, him, her, it, us, you, them
ငါကိုယ်တိုင်၊ ကျုပ်ကိုယ်တိုင်၊ မင်းကိုယ်တိုင်၊ သားကိုယ်တိုင်၊ သမီးကိုယ်တိုင်၊ မင်းတို့ကိုယ်တိုင်၊ ကျွန်တော့်ဘာသာ၊ ကျုပ်ဘာသာ၊ ငါ့ဘာသာသာ၊ မိမိဘာသာ၊ သူ့ဘာသာ၊ မင်းဘာသာ၊ သားဘာသာ၊ သမီးဘာသာ၊ မင်းတို့ဘာသာ၊ သားတို့ဘာသာ၊ သမီးတို့ဘာသာ၊ ကျွန်တော်တို့ဘာသာ၊ သူကိုယ်တိုင်၊ ကျုပ်တို့ကိုယ်တိုင်၊ သူမကိုယ်တိုင်၊ သူ့ကိုယ်ကို၊ ကိုယ့်ကိုကိုယ်၊ ငါ့ကိုယ်ကို၊ ကျုပ်ကိုယ်ကို၊ မိမိကိုယ်ကို၊ ၎င်းပင်၊ ထိုအရာပင်	myself, yourself, himself, herself, itself, ourselves, yourselves, themselves, oneself
သည့်၊ မည့်၊ တဲ့၊ အဲဒါ	that
ကျွန်ုပ်၏၊ ကျွန်တော်၏၊ ကျုပ်၏၊ ကျွန်မ၏၊ ကျနော်၏၊ ကျမ၏၊ သူ၏။ သူမ၏။ ထိုအရာ၏၊ ထိုဟာ၏၊ ငါ၏၊ ငါတို့၏၊ ကျွန်ုပ်တို့၏၊ ကျွန်တော်တို့၏၊ ကျွန်မတို့၏၊ ကျနော်တို့၏၊ ကျမ၏၊ သင်၏၊ သင်တို့၏၊ မင်း၏၊ မင်းတို့၏၊ သူတို့၏၊ ကျွန်တော့်ဟာ၊ ငါ့ဟာ၊ သူ့ဟာ၊ သူမဟာ၊ ကျနော်ဟာ၊ ကျမဟာ၊ ကျနော်တို့ဟာ၊ ကျမတို့ဟာ၊ သူတို့ဟာ၊ မင်းတို့ဟာ၊ သားဟာ၊ သမီးဟာ၊ သားတို့ဟာ၊ သမီးတို့ဟာ၊	my, your, his, her, its, our, your, their, mine, yours, his, hers, ours, yours, theirs

(continued)

Table 2 (continued)

ဟောဒါ၊ ဟောဒီ၊ ထိုအရာ၊ ၎င်းအရာ၊ ယင်းအရာ၊ အဲဒါ၊ အဲဒါတွေ၊	this, that, these, those
အချို့၊ တစ်ခုခု၊ အဘယ်မဆို၊ ဘယ်အရာမဆို၊ အရာရာတိုင်း၊ ဘယ်လောက်မဆို၊ တစုံတရာ၊ အလျဉ်းမဟုတ်၊ မည်သည့်နည်းနှင့်မျှမဟုတ်၊ အားလုံး၊ အလုံးစုံ၊ အရာခပ်သိမ်း၊ တယောက်စီ၊ တခုစီ၊ အကုန်၊ အပြည့်အစုံ၊ လုံးလုံး၊ နှစ်ခုလုံး၊ နှစ်ယောက်လုံး၊ တစ်စုံတရာ၊ တစုံတခု၊ တခုခု၊ တယောက်ယောက်၊ ဘယ်သူမဆို၊ မည်သူမဆို ဘယ်သူမှ၊ လူတိုင်း၊ လူတကာ၊ ဘယ်အရာမျှမရှိ၊	some, any, no, none, other, another, every, all, others, each, whole, both, neither, someone, somebody, something, anyone, anybody, anything, nobody, nothing, everyone, everybody, everything,
နှင့်၊ ပြီးလျှင်၊ အဲဒီနောက်၊ သို့မဟုတ်၊ ဒါမှမဟုတ်၊ ဖြစ်စေ၊ ဒါပေမဲ့၊ ဒါပေမယ့်၊ မှတပါး၊ မှလွဲလျှင်၊ အဘယ်ကြောင့်ဆိုသော်၊ ဘာကြောင့်လဲဆိုတော့၊ ဒါကြောင့်၊ အဲဒါကြောင့်၊ သဖြင့်၊ ၍၊ သည့်အတွက်ကြောင့်၊ လျှင်၊ ပါက၊ အကယ်လို့၊ အကယ်၍၊ သော်ငြားလည်း၊ စေကာမူ၊ နည်းတူ၊ ပေမဲ့၊ ထိုနည်းတူစွာ၊ အဲလိုပဲ၊ ကဲ့သို့၊ နှင့်စပ်လျဉ်း၍၊ ယင်းကဲ့သို့၊ ၎င်းကဲ့သို့၊ ဒါလောက်တော့၊ ဒါကတော့၊ အခုလောက်ထိ၊ ဤမျှလောက်၊ ဒါလောက်ကတော့၊	and, or, but, because, if, as, such
မည်ကဲ့သို့၊ မည်သည့်နည်းနှင့်၊ မည်သည့်နည်းဖြင့်၊ မည်သို့၊ ဘယ်လိုလဲ၊ သို့ပေမည့်၊ မည်သည့်နည်းနဲ့မဆို၊ မည်သည့်နည်းနှင့်မဆို၊ ဘာနည်းနဲ့၊ ဘယ်နည်းနှင့်၊ မည်ရွေ့မည်မျှ၊ အဘယ်မျှလောက်၊ ဘယ်လောက်၊ ဘယ်သူ၊ မည်သူ၊ မည်သည့်အကြောင်းကြောင့်၊ ဘယ်လိုကြောင့်၊ ဘာအတွက်နဲ့၊ ဘာကြောင့်၊ ဘယ်သူကြောင့်၊ ဘယ်နေရာ၊ ဘယ်ကို၊ ဘယ်သူ့ကို၊ အဘယ်ကြောင့်၊ မည်သည့်၊ ဘာလဲ၊ ဘယ်လဲ၊ ဘယ်သူ၏၊ ဘယ်နားမှာ၊ ဘယ်နေရာမှာ၊ ဘယ်အချိန်၊ ဘယ်အခါ၊ မည်သည့်အချိန်၊ ဘယ်တော့၊ မည်သူ့ကို၊ ဘယ်သူ့ကို၊ မည်သူ့ကို၊ မည်သည့်အရာ	how, who, why, what, where, whose, when, whom, which
ဘယ်လိုပဲဖြစ်ဖြစ်၊ ဘာပဲဖြစ်ဖြစ်၊ ဘယ်သူနဲ့ဖြစ်စေ၊ မည်သူမဆို၊ ဘယ်သူမဆို၊ အဘယ်သူမဆို၊ မည်သည့်အရာမဆို၊ ဘာဖြစ်ဖြစ်၊ ဘယ်နေရာပဲဖြစ်ဖြစ်၊ အချိန်မရွေး၊ နေရာမရွေး၊ ဘယ်အချိန်ပဲဖြစ်ဖြစ်၊ မည်သို့ပင်ဖြစ်စေ၊	however, whoever, whatever, wherever, whenever, whomever,

Example 1 Segmented Result

Text in original status:	အခုတော့သာသာယာယာပဲ (Now I am happy)
Segmented result:	အခု_တော့_သာသာယာယာ_ပဲ

The segmented sentence may contain unnecessary stop words. Therefore, the stop words searching and removing process is performed as third-step. Example 2 shows the stop words removal results of a status.

Example 2 Stop Words Removal Result

Segmented result:	အခု_တော့_သာသာယာယာ_ပဲ
Stop words:	အခု, တော့, ပဲ
Extract Words (Terms)	သာသာယာယာ (happy)

The segmentation and stop words removal process is performed on all filtered status. After preprocessing, the words or terms to find the associated emotions are got. The lexicon creation using the gathering words will be discussed in next section.

4.3 Lexicon Creation

In the lexicon creation phase, M-Lexicon is created by using matrices and normalization as shown in Fig. 1. The Facebook status as in Table 3, are preprocessed to get the terms or words.

This system maps the six Facebook reactions into corresponding emotion as mentioned in Table 4. Number of Facebook reaction is the number of the corresponding emotion. The values can be obtained from Facebook reactions, fear and disgust. 'Like' reaction is not associated with any emotion in lexicon creation. The count for 'Disgust' and 'Fear' reaction is additionally collected from Facebook user as voting. Table 5 describes the reaction-emotion mapping of sample status (in Table 3).

After the original status is pre-processed, the system gets word which to be future used as words in the lexicon. Word-by-status matrix is a matrix that contains words from preprocess phase as rows, and the columns be status. In order to get the associate value of an emotion, TF-IDF scheme as in Eq. (1) is applied.

$$\text{tfidf}(t, s, S) = \text{tf}(t, s) \times \text{idf}(t, S), \tag{1}$$

where t denotes the terms, s denotes each status, and S denotes the collection of status. Term Frequency (tf) measures how frequently a term occurs in a status. The

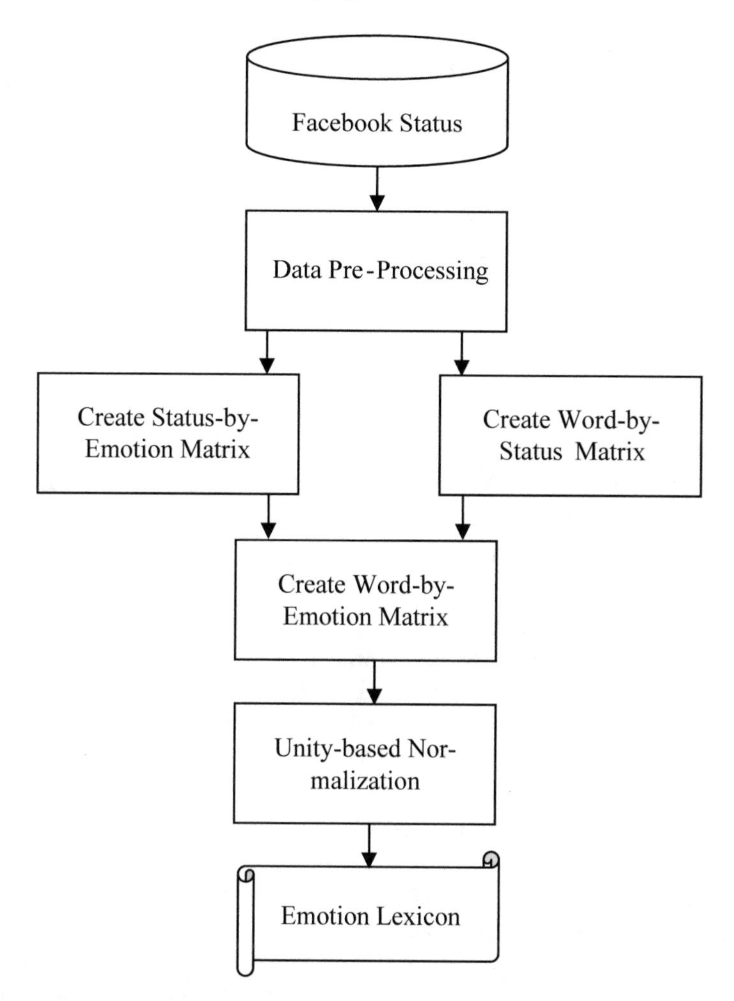

Fig. 1 M-Lexicon creation

Table 3 Sample status

Status#	Status Body
Status1	စိတ်ညစ်စရာတွေနဲ့ကြုံခဲ့ရတယ် (I was sad)
Status2	အခုတော့သာသာယာယာပဲ (Now I am happy)

term frequency (tf) is often divided by the status length (the total number of terms in the status) as a way of normalization:

$$\text{tf}(t, s) = \frac{\text{Number of times term t apperars in a status s}}{\text{Total number of terms in the status s}} \tag{2}$$

Table 4 Reaction-emotion mapping

Facebook reactions	Emotions
Like	–
Love	Happiness
Ha Ha	Happiness
Wow	Surprise
Sad	Sadness
Angry	Anger

Inversed document frequency (idf) is a measure of how much information the word provides, that is, whether the term is common or rare across all status. The value is obtained by dividing the total number of status by the number of status containing the term, and then taking the logarithm of that quotient as in Eq. (3).

$$\mathrm{idf}(\mathrm{t}, \mathrm{S}) = \log \frac{\text{Total number of status S}}{\text{Number of status with terms t in it}} \quad (3)$$

Table 6 depicts a sample of word-by-status matrix.

The status-by-emotion matrix, as shown in Table 7, is a simply matrix which contains each status for rows and six basic emotions are columns. In order to get affective annotation value, the system firstly sums each reaction and voted values. Then, it calculates average percentage of each emotion using Eq. (4).

$$\mathrm{avg}(\mathrm{emotion}, \mathrm{status}) = \frac{\mathrm{NE}}{\mathrm{NU}} \times 100\%, \quad (4)$$

where NE denotes the number of voted or reacted for each emotion, NU is the number of voted or reacted user for each emotion.

The word-by-emotion matrix is a matrix, which contains each word from word-by-status as rows and six emotions as columns. The value of word-by-emotion matrix

Table 5 Reaction-emotion mapping of sample status

#	Status	Happiness	Sadness	Fear	Anger	Surprise	Disgust
1	Status 1	10	1	0	0	11	0
2	Status 2	11	5	5	0	0	0

Table 6 Word-by-status matrix

#	Words	Status1	Status2
1	စိတ်ညစ် (sad)	0.03	0.0
2	သာသာယာယာ(happy)	0.0	0.06

Table 7 Status-by-emotion matrix

#	Status	Happiness	Sadness	Fear	Anger	Surprise	Disgust
1	Status 1	0.05	0.5	0.0	0.0	0.0	0.0
2	Status 2	0.48	0.0	0.0	0.0	0.0	0.0

is obtained by multiplying the word-by-status matrix with status-by-emotion matrix. Table 8 shows a sample of word-by-emotion matrix.

The values obtained in word-by-emotion matrix may vary. In M-Lexicon, the system defines only two states for each word. The first one is that if a word is an emotion, the value is set as 1. The last one is that if the word is not associated with an emotion, the value is set as 0 in the emotion. Thus, the system performs the normalization process on the value of word-by-emotion matrix to adjust it in range [0, 1]. As shown in Eq. (5), min-max or unity-based normalization is applied. The value generalized to restrict the range of values in the dataset between any arbitrary points: new_max and new_min.

$$v' = \frac{v - \min_A}{\max_A - \min_A}(\text{new\_max}_A - \text{new\_min}_A) + \text{new\_min}_A, \quad (5)$$

where new_min is the minimum of the normalized dataset and new_max is the maximum of the normalized dataset. v is old variable. v' is transformed variable. Table 9 listed the normalized result on the word-by-emotion matrix. Table 10 illustrates sample of M-Lexicon.

Table 8 Word-by-emotion matrix

#	Words	Happiness	Sadness	Fear	Anger	Surprise	Disgust
1	စိတ်ညစ် (sad)	0.0015	0.015	0.0	0.0	0.0	0.0
2	သာသာယာယာ (happy)	0.0288	0.0	0.0	0.0	0.0	0.0

Table 9 Normalized emotion lexicon

#	Words	Happiness	Sadness	Fear	Anger	Surprise	Disgust
1	စိတ်ညစ် (sad)	0	1	0	0	0	0
2	သာသာယာယာ (happy)	1	0	0	0	0	0

Table 10 M-Lexicon sample data

#	Words	Happiness	Sadness	Fear	Anger	Surprise	Disgust
1	ဝမ်းသာ (joy)	1	0	0	0	0	0
2	ဒေါကန် (angry)	0	0	0	1	0	0
3	သာသာယာယာ (happy)	1	0	0	0	0	0
4	ထိတ်လန့် (afraid)	0	0	1	0	0	0
5	မုန်းတီး (hate)	0	0	0	0	0	1
6	အံ့အားသင့် (amazing)	0	0	0	0	1	0
7	စိတ်ညစ် (sad)	0	1	0	0	0	0
8	စိတ်တို (angry)	0	0	0	1	0	0
9	ကြည်နူး (delightful)	1	0	0	0	0	0
10	စိတ်အားငယ် (depressive)	0	1	0	0	0	0

The results are used to insert into or update to the existing word in M-Lexicon as the following conditions:
Condition 1 (Update existing word): If a word already exists in M-Lexicon, the system updates the word-associated value in each emotion. If the existing value is 0 and the new value is 1, the existing value will be overwritten by new value. For instance, the Myanmar word 'မုန်းတီး'(hate) already exists in M-Lexicon with 1 in disgust and 0 in other emotions. That word with new values: 1 in anger and 0 in others. The existing value is updated as 0 with 1 in anger emotion. If the existing value is 1 and the new value is 0, the update process will not be performed. For example, the word 'သာသာယာယာ'(happy) is in M-Lexicon with 1 in happiness and 0 in other emotion. That word has new values as 0 in all emotions. The existing 1(happiness) is not updated into 0.
Condition 2 (Insert new word): If the word does not exist in M-Lexicon, then the new value with at least one associated emotion is inserted into lexicon as new word. If the word is not associated with any emotion which means all emotion values are 0,

the words will not be inserted into M-Lexicon. For example, the word 'စိတ်ညစ်'(sad) is not in M-Lexicon, the word has 1 in sadness and 0 in other emotions. The word 'စိတ်ညစ်'(sad) is inserted into M-Lexicon as new word. Although the word 'စိတ်'(heart) does not exist in lexicon, it will not be inserted into lexicon because it has 0 value in all emotions.

5 Experiment

We initially create M-Lexicon by using 485 Facebook status with 3890 Myanmar words. After passing the pre-processing phase, there are 2147 words with 1743 stop words. The 2147 words are analyzed into corresponding emotions. Finally, we got 1947 words for each six emotions as shown in Table 11.

For each word in initial lexicon, we use the manual emotion labeling by human to compute the accuracy. The accuracy measurement of M-Lexicon is done in terms of precision, recall, and F-Measure.

The accuracy measurement of M-Lexicon is done in terms of precision, recall, and F-Measure.

$$\text{Precision} = \frac{\text{TP}}{(\text{TP} + \text{FP})} \tag{6}$$

TP True Positive
FP False Positive

Precision is the number of correctly associated emotion labeled words in lexicon divided by all associated emotion words in lexicon.

$$\text{Recall} = \frac{\text{TP}}{(\text{TP} + \text{FN})} \tag{7}$$

FN False Negative

Recall is the number of correctly associated emotion labeled words in lexicon divided by words annotated as correct.

Table 11 Emotions in M-Lexicon

Emotions	#Number
Happiness	800
Sadness	600
Fear	124
Anger	133
Surprise	155
Disgust	135

Table 12 Accuracy results

Emotions	Precision	Recall	F-measure
Happiness	0.826	0.811	0.818
Sadness	0.790	0.778	0.784
Fear	0.532	0.512	0.522
Anger	0.827	0.797	0.812
Surprise	0.822	0.809	0.815
Disgust	0.6	0.6	0.6

$$F\text{-}measure = 2 \times \frac{Precision \times Recall}{Precision + Recall} \tag{8}$$

After precision and recall are calculated, the values are used to calculate the f-measure, the harmonic mean of precision and recall that functions as a weighted average equation.

Table 12 shows the results of the accuracy measurement on M-Lexicon. The system currently obtains average 73% f-measure for six basic emotions which especially for Myanmar language.

6 Conclusion

In this paper, the creation of a word-emotion lexicon called M-Lexicon was described. M-Lexicon was generated by using Facebook users' status. Words in preprocessed status were used as lexicon words, the associated emotion value was calculated by using matrices, and finally the outcome values were normalized to adjust them in range [0, 1]. Currently, the lexicon is created on 485 Facebook users' status that has total 3890 words and the lexicon has 1947 words. This paper also measured the accuracy of M-Lexicon. The current experiment shown that lexicon contains 73% of corrected emotion words.

M-Lexicon is currently in initial creation that is to be future used in emotion detection system for Social Media (Facebook) users in Myanmar language. To better and more emotion words, we will continually build emotion lexicon with newly updated several status and additional stop words.

References

1. Ekman, P.: An argument for basic emotions. Cogn. Emot. **6**(3), 169–200 (1992)
2. Ekman, P.: Emotion in the Human Face. Oxford University Press (2005)
3. Ekman, P., Friesen, W.V.: Unmasking the Face: A Guide to Recognizing Emotions From Facial Expressions. Marlor Books (2003)

4. Bandhakavi, A., Wiratunga, N., Deepak, P., Massie, S.: Generating a word-emotion lexicon from #emotional tweets. In: Proceedings of the Third Joint Conference on Lexical and Computational Semantics (*SEM2014), Dublin, Ireland, pp. 12–21 (2014, August)
5. Xu, G., Meng, X., Wang, H.: Build Chinese emotion lexicons using a graph-based algorithm and multiple resources. In: Proceedings of the 23rd International Conference on Computational Linguistics (Coling 2010), Beijing, China, pp. 1209–1217 (2010, August)
6. Do, H.J., Choi, H.: Korean twitter emotion classification using automatically built emotion lexicons and fine-grained features. In: 29th Pacific Asia Conference on Language, Information and Computation, Shanghai, China, pp. 142–150 (2015, October–November)
7. Mohammad, S.M., Turney, P.D.: Crowdsourcing a word-emotion association lexicon. Comput. Intell. **29**(3), 436–465 (2013)
8. Plutchik, R.: The nature of emotions human emotions have deep evolutionary roots, a fact that may explain their complexity and provide tools for clinical practice. Am. Sci. **89**(4), 344–350 (2001)
9. Myanmar Language Commission: Myanmar Dictionary, 2nd edn. University Press (2008)
10. Myanmar Language Commission: Myanmar-English Dictionary, 11th edn. University Press (2011)
11. Thet, T.T., Na, J., Ko, W.K.: Word segmentation for the Myanmar language. J. Inf. Sci. **34**(5), 688–704 (2008)
12. Maung, Z.M., Makami, Y.: A rule-based syllable segmentation of Myanmar text. In: IJCNLP-08 Workshop of NLP for Less Privileged Language, Hyderabad, India, pp. 51–58 (2008, January)
13. Hopple, P.: The structure of nominalization in burmese. Ph.D. thesis, 2003
14. Word Breaker: https://flask-py-word-breaker.herokuapp.com

Release from the Curse of High Dimensional Data Analysis

Shuichi Shinmura

Abstract Golub et al. started their research to find oncogenes and new cancer subclasses from microarray around 1970. They opened their microarray on the Internet. The other five medical projects published their papers and released their microarrays, also. However, because Japanese cancer specialist advised us that NIH decided that these researches were useless after 2004, we guess medical groups abandoned these researches. Although we are looking for NIH's report, we cannot find it now. Meanwhile, many researchers of statistics, machine learning and bioengineering continue to research as a new theme of high-dimensional data analysis using microarrays. However, they could not succeed in cancer gene analysis as same as medical researches (Problem5). We discriminated six microarrays by Revised IP-OLDF (RIP) and solved Problem5 within 54 days until December 20, 2015. We obtained the two surprising results. First, MNMs of six microarrays are zero (Fact3). Second, RIP could decompose microarray into many linearly separable gene subspaces (SMs) and noise subspace (Fact4). These two new facts indicate that we are free from the curse of high dimensional microarray data and complete the cancer gene analysis. Because all SMs are LSD and small samples, we thought to analysis all SMs by statistical methods and obtained useful results. However, we were disappointed that statistical methods do not show linearly separable facts and are useless for cancer gene diagnosis (Problem6). After trial and error, we make signal data made by RIP discriminant scores (RipDSs) from SM. Through this breakthrough, we find useful information by correlation analysis, cluster analysis, and PCA in addition to RIP, Revised LP-OLDF and hard margin SVM (H-SVM). We think that the discovery of the above two new facts is the essence of Problem5. Moreover, we claim to solve Prpblem6 and obtain useful medical care information from signal data as a cancer gene diagnosis. However, our claim needs validation by medical specialists. In this research, we introduce the reason why no researchers could succeed in the cancer gene diagnosis by microarrays from 1970.

S. Shinmura (✉)
Emeritus Seikei University, 1-8-7-301 Sakasai Kashiwa City, Chiba 277-0042, Japan
e-mail: sshinmura2@gmail.com

R. Lee (ed.), *Big Data, Cloud Computing, and Data Science Engineering*,
Studies in Computational Intelligence 844,
https://doi.org/10.1007/978-3-030-24405-7_12

Keywords Minimum number of misclassifications (MNM) · Linearly separable data (LSD) · Cancer gene diagnosis

1 Introduction

Golub et al. [15] started their research to find oncogenes and new cancer sub-classes from microarray around 1970. They opened their microarray. The other five US medical projects published their papers and released their microarrays from 1999 to 2004. These six microarrays consist of cancer and health classes, or two different types of cancers. Because many researchers of statistics, machine learning, pattern recognition, and bioengineering can use high dimensional data in free charge, they studied this research as a new theme of big data. However, they pointed out three difficulties of cancer gene analysis. These difficulties showed they could not succeed in their researches to specify oncogenes from microarrays as same as the medical projects (Problem5). Because the statistical discriminant functions are useless for cancer gene analysis, they could not succeed in Problem5 since 1970.

On the other hand, we found four serious problems of discriminant analysis through many discriminant studies. Problem1 is the defect of the number of misclassifications (NM). Problem2 is linearly separable data (LSD) discrimination. Problem3 is the defect of variance-covariance matrices. Problem4 is that discriminant analysis is not the inferential statistics. We established the new theory of discriminant analysis in 2015 [23–29]. We developed four mathematical programming (MP) based linear discriminant functions (LDFs) and the 100-fold cross validation for a small sample method (Method1) in addition to finding two new facts. We solved four problems by the theory in 2015. Especially, to solve Problem1 and Problem2 relate to the success of cancer gene analysis.

After establishing the theory, we knew many researchers in addition to medical projects studied the high-dimensional microarray data analysis as a new theme. However, they could not succeed in this research (Problem5). We downloaded six microarrays [17] on October 28, 2015. We discriminated these data as an applied problem of the theory and solved Problem5 within 54 days until December 20, 2015. We obtained the two surprising results. First, minimum NMs (MNMs) of six microarrays are zero and LSD (Fact3). Second, Revised IP-Optimal LDF (Revised IP-OLDF, RIP) based on MNM criterion could decompose each microarray into many linearly separable gene subspaces (Small Matryoshkas, SMs) and noise subspace (Fact4). These two new facts indicate that we are free from the curse of high dimensional microarray data and complete the cancer gene analysis by MP.

Because all SMs are LSD and small samples (small n and small p), we thought to analysis all SMs by statistical methods and obtained useful results. However, statistical methods did not show linearly separable facts and were useless for cancer gene diagnosis (Problem6). After trial and error, we find that many RatioSVs by RIP discriminant scores (RIP DSs, RipDSs) of SMs are over 5%, and we consider these RipDSs are malignancy indexes for cancer gene diagnosis. Moreover, we make signal data made by these RipDSs instead of genes included in SMs. Through this breakthrough, we find useful information by t-test, correlation analysis, cluster analysis, and PCA in addition to RipDSs, Revised LP-OLDF discriminant scores (LpDSs) and hard margin SVM (H-SVM) discriminant scores (HsvmDSs). We think that the discovery of the above two new facts is the essence of Problem5. Moreover, we claim to solve Problem6 and obtain useful medical care information from signal data as a cancer gene diagnosis [34]. However, our claim must be validated by medical specialists because our results are medical matters.

We explained our cancer gene diagnosis to a medical specialist by our results [30] in 2017. He told us NIH judged these researches were useless for Problem5 and Problem6 except for the breast cancer. He advised us not to continue our research because physicians stopped their researches world-widely and were not interested in our results. However, we ignored his advice because of the following four reasons:

(1) Although the discriminant analysis is the most proper method for Poblem5, no researchers could succeed because the existing discriminant functions were useless for Problem5. Researchers do not understand the defect of NM (Problem1) and the relation of NM and discriminant coefficients (Fact1). These understandings are essential for high-dimensional data analysis.
(2) Although microarrays are LSD, there was no research on LSD-discrimination except for us (Problem2). Thus, there were no researches pointed out that six microarrays are LSD. However, Aoshima and Yata [2, 3] approached many microarrays including six microarrays by high-dimensional classifiers. They showed that two classes locate on two different high dimensional balls for several datasets (Fact3).
(3) Because only RIP and H-SVM [40] can discriminate LSD theoretically, both LDFs can discriminate microarrays correctly. Although some medical researches discriminate microarrays by SVM, it is bizarre that there were no papers to point out Fact3. Probably, they used soft-margin SVM such as SVM1 (penalty $c = 1$). Or their solver could not solve Problem5 correctly. If the neighborhoods of the optimal solutions are flat, we consider their quadratic programming (QP) solvers may not converge to the correct solution described in Fig. 1.
(4) Only RIP defined by integer programming (IP) and Revised LP-OLDF defined by linear programming (LP) can decompose microarrays into many SMs and noise gene subspace (MNM > 0) quickly. We explain the reason why H-SVM defined by QP cannot decompose microarray into SMs [31–33, 35].

Although many researchers know how to obtain the solution of simultaneous equations and the optimal solution of LP, no one has ever unified both pieces of knowledge. Moreover, to think of discriminant theory in the small world of statistics is the cause of failure. Because we are free from the curse of high-dimensional data by RIP and Revised LP-OLDF, everyone can analyze all SMs by the statistical methods and contribute cancer gene diagnosis. We proposed the standard procedure on how to analyze all SMs. In this research, we introduce the reason why no researchers could succeed in Problem5 and Problem6 from 1970, also.

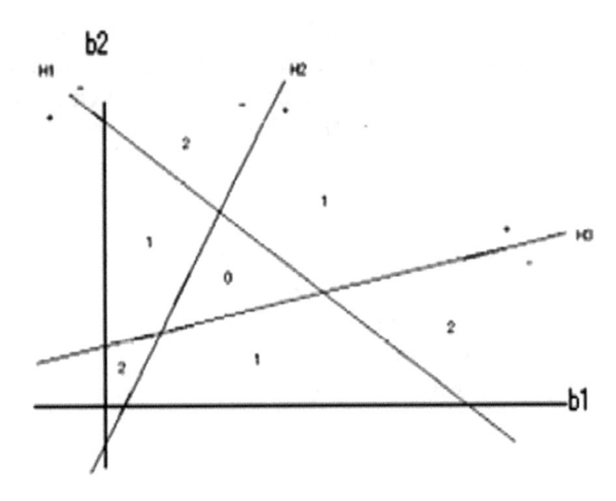

Fig. 1 The relation of NM and LDF coefficient

2 New Theory of Discriminant Analysis

We found four severe problems with discriminant analysis. These problems were solved by our theory that consists of four OLDFs and two new facts. Furthermore, the Matryoshka feature selection method (Method2) achieved by LINGO could quickly solve Problem5 as an applied problem. LINGO is the mathematical programming (MP) solver of linear programming (LP), quadratic programming (QP), integer programming (IP), nonlinear programming (NLP), and stochastic programming with model building functions developed by LINDO Systems Inc. An evaluation version of LINGO, What's Best! (Excel add-in software) and LINDO API (c library to develop the optimization systems) are downloaded free from the HP. Manuals, textbook [22], and many sample models are downloaded free from the HP (https://www.lindo.com/). Our LINGO programs are explained in [29] or [34].

2.1 *The Outlook of the Theory and Five Problems*

Because we had already established the theory [29], we overview this research. By many experiences through discriminant analysis, we found four severe problems in addition to two facts [27, 28]. If researchers discriminate real data, they can easily find these problems. However, nobody pointed out these problems. The reason is that even if the actual research subject is not a normal distribution, they are studying in a restricted world assuming a normal distribution. Because we solved four problems, we tried the Problem5 as applied research of the theory and solved this problem within 54 days in 2015. In this paper, we introduce the cancer gene diagnosis by microarrays. However, researchers must understand MP in addition to statistics, and challenge the analysis of real data. We find five problems of discriminant theory as follows.

2.1.1 Problem1: Defect of NM Solved by Fact1

Let us consider two-class discrimination that consists of n cases (n_1 cases belong to class1, and n_2 cases belong to class2) and p-independent variables $\mathbf{x}$. The $y_i = 1$ for class1 and $y_i = -1$ for class2 for each $\mathbf{x}_i$. Equation (1) defines LDF such as $f(\mathbf{x})$. The $\mathbf{b}$ is p-dimensional discriminant coefficients. The b_0 is the intercept.

$$f(\mathbf{x}) = \mathbf{b} * \mathbf{x} + b_0 \tag{1}$$

Most researchers erroneously believe the discriminant rule is as follows: If $f(\mathbf{x}_i) \geq 0$, $\mathbf{x}_i$ belongs to class1. If $f(\mathbf{x}_i) < 0$, $\mathbf{x}_i$ belongs to class2. This understanding is completely wrong. The true discriminant rule is as follows:

(1) If $y_i{*}f(\mathbf{x}_i) > 0$, $\mathbf{x}_i$ is correctly classified to the class1 or class2.
(2) If $y_i{*}f(\mathbf{x}_i) < 0$, $\mathbf{x}_i$ is misclassified to the class1 or class2.
(3) If $y_i{*}f(\mathbf{x}_i) = 0$, we cannot decide which class $\mathbf{x}_i$ belongs until now.

Although Problem1 was the unresolved problem, the notation of IP-OLDF found the relation of NMs and LDF coefficients (Fact1) and solved Problem1 [24]. There are many defects of NM as follows:

(1) Different LDFs such as Fisher's LDF [12], logistic regression [9, 11], regularized discriminant analysis (RDA) [14], four OLDFs and three SVMs give us different NMs.
(2) If we change the discriminant hyperplane, we obtain different NMs.
(3) If k cases are on the discriminant hyperplane ($f(\mathbf{x}_i) = 0$), then the correct NM may increase to (k + NM).

Although our claim is logical and simple, no one has pointed out strangely.

On the other hand, Miyake and Shinmura developed a heuristic optimal LDF (OLDF) based on MNM criterion after the investigation of an error rate of LDF [19]. However, statistical journals rejected our paper because of MNM standard overfitted and overestimated the sample. Medical engineering journal accepted our paper after over four years [20]. MNM is the most important statistics for LSD-discrimination instead of NM.

2.1.2 Problem2: LSD-Discrimination

Vapnik [40] defines LSD-discrimination clearly. We modify his idea as follows:

(1) If $y_i{*}f(\mathbf{x}_i) \geq 1$, $\mathbf{x}_i$ is correctly classified the class1 or class2.
(2) There is no case in the range of $-1 < f(\mathbf{x}_i) < 1$.

We consider two support vector SV = ± 1 instead of discriminant hyperplane f(**x**) = 0. This definition is better than the above true discriminant rule. We firstly defined IP-OLDF. However, it cannot find true optimal convex polyhedron (OCP) introduced in Fig. 1 if the data does not satisfy the general position. Thus, we developed a RIP that looks for the inner point directly. However, two referees in Japanese and European journals rejected our papers because LSD-discrimination was easy, and the purpose of the discriminant analysis was to discriminate the overlapping data. However, they are not logical because they did not know the statistics MNM. We can firstly define MNM = 0 equal to LSD. They do not know MNM that clearly defines LSD (MNM = 0) and overlapping data (MNM > 0). Also, nobody pointed out that microarrays are LSD, as the importance of LSD discrimination studies is not known. This fact is the first reason why Problem5 was not solved because of the fundamental knowledge of discriminant theory has been ignored until now.

2.1.3 Problem3: The Defect of Generalized Inverse Matrix

The pass/fail judgment of the exam is the self-evident LSD-discrimination. However, when Fisher's LDF discriminated the research data of the university entrance examination center in Japan, the error rate was over 20% in mathematics [26]. Furthermore, when the passing score is 90% of the score total, the quadratic discriminant function (QDF) and RDA misclassify all the pass candidates. The reason is caused by results that all successful applicants answered some questions correctly. This result is caused by the defect of the generalized inverse matrix. If we add a random number to a constant value, we can quickly solve it. We wasted three research years in the wrong approach.

2.1.4 Problem4: Inferential Statistics

Fisher never formulated the standard error of error rate and discriminant coefficients. Therefore, we developed the 100-fold cross validation for small sample (Method1) and the best model as model selection instead of Leave-one-out (LOO) method [18]. We calculate two minimum average of error rates M1 and M2 for the 100 training samples and 100 validation samples. Because MNM decreases monotonously (Fact2), the error rate of the full model always becomes M1. Thus, M1 is not proper for model selection. M2 of RIP is better than other seven LDFs such as two other OLDFs, three SVMs, Fisher's LDF and logistic regression using six different types of ordinary data. This truth indicates that MNM criterion is free from overfitting. Although many medical researchers validated their results of Problem5 by LOO that was developed in the age of weak computing power, we consider the no need for validation of our results because two classes are separable completely. We explain this matter using RatioSV that is the other important statistics of LSD-discrimination later.

2.1.5 Problem5: Cancer Gene Analysis by Microarray

Golub et al. [15] published their paper at Science in 1999 and confessed that they started their research about 30 years ago. Thus, we judged that they started their study around 1970. Probably, Prof. Golub is the pioneer of Problem5. Because they developed several unique methods instead of statistical methods, we think they decided most statistical methods were useless for Problem5, except for cluster analysis. They should have complained to engineering researchers, especially discriminant theorists including us, complaining of the consequences of their failure.

We consider other non-medical researchers studied this theme after around 1990 because the commercial microarray equipment was used after 1990. Probably, they did not know the NIH's decision. They published many papers about feature selection and filtering system to separate signal and noise from the high dimensional microarray. However, no researchers solved Problem5 entirely. In this research, we explain the reason why they could not succeed in the cancer gene diagnosis (Problem6) in addition to cancer gene analysis (Problem5). No researchers found Fact3. This theme is Problem5. Next, we developed LINGO Program3 to solve Method2 by three OLDFs and three SVMs [34]. RIP and Revised LP-OLDF can decompose microarray into many SMs (MNM $= 0$) and the noise gene subspace (MNM > 0). This truth is Fact4. Only H-SVM and two OLDFs can find Fact3. However, only RIP, and Revised LP-OLDF can find Fact4. By this breakthrough, because we are free from the curse of high dimensional data, we can propose the cancer gene diagnosis by JMP. We are not the specialists of this area. However, because the fact of LSD is a critical signal of cancer gene diagnosis, our results are precise for everyone. Only the cooperation of MP and statistics can succeed in the cancer gene analysis and diagnosis. Notably, LINGO [22] and JMP [21] enhanced our intellectual productivities because we introduced LINDO, LINGO, SAS and JMP into Japan and published many books and papers.

2.2 *IP-OLDF and Two New Facts*

For the explanation of IP-OLDF, we used Fig. 1 in 1997. Later, we found two new facts by Fig. 1. When we downloaded six microarrays [17] on October 28, 2015, and discriminated Shipp microarray [36]. RIP, Revised LP-OLDF, and H-SVM found Fact3. By checking the coefficients, we found that the number of non-zero coefficients was less than n. On the other hand, most coefficients of H-SVM were not zero. Thus, we developed the LINGO Program3 that achieved Method2. Program3 of RIP found all SMs of six microarrays in Table 1 until December 20, 2015.

Table 1 Six Microarrays and SMs [29]

Data	Alon et al. [1]	Chiaretti et al. [7]	Tian et al. [39]
Size	62 * 2,000	128 * 12,625	173 * 12,625
SM: Gene	64: 1,999	270: 5,385	159: 7,221
NM/error rate	52/8%	55/2%	88/18%
Data	Golub et al. [15]	Shipp et al. [36]	Singh et al. [37]
Size	72 * 7,129	77 * 7,129	102 * 12,625
SM: Gene	69: 1,238	213: 3,032	179: 3,990
NM/error rate	57/11%	51/4%	45/10%

2.2.1 IP-OLDF Based on MNM Criterion

IP defines IP-OLDF based on MNM criterion in (2). The e_i is 0/1 integer variable corresponding to classified cases (0) or misclassified cases (1). We fix the intercept of IP-OLDF at 1 and define it in p-dimensional coefficient space. Because pattern recognition researchers define the intercept is free variable, it is in (p + 1) dimension. They could not explain the relations between NM and LDF coefficient more in-depth.

$$\mathrm{MIN} = \sum e_i;\ y_i * \left({}^t\mathbf{x}_i * \mathbf{b} + 1\right) \geq -e_i; \tag{2}$$

Although $y_i*({}^t\mathbf{x}_i*\mathbf{b} + 1)$ is extended discriminant scores (DS), ${}^t\mathbf{x}_i*\mathbf{b} + 1 = 0$ is a linear hyperplane that divides p-dimensional discriminant coefficient space to two half-planes such as plus half plane ($y_i*({}^t\mathbf{x}_i*\mathbf{b} + 1) > 0$) and minus half plane ($y_i*({}^t\mathbf{x}_i*\mathbf{b} + 1) < 0$). If we choose **b** in plus hyperplane as LDF, LDF discriminate $\mathbf{x}_i$ correctly because of $y_i*({}^t\mathbf{x}_i*\mathbf{b} + 1) = y_i*({}^t\mathbf{b}*\mathbf{x}_i + 1) > 0$. On the other hand, if we choose **b** in a minus hyperplane, LDF misclassifies $\mathbf{x}_i$ because of $y_i*({}^t\mathbf{x}_i*\mathbf{b} + 1) = y_i*({}^t\mathbf{b}*\mathbf{x}_i + 1) < 0$. However, we must solve the other two models such as the intercept = −1 and 0. It looks for the right vertex of an OCP if data is a general position. There are only p-cases on the discriminant hyperplane, and it becomes the vertex of true OCP. We find two relevant facts as follows.

2.2.2 Two Fact1

IP-OLDF is a flawed OLDF affected by Problem1, but we find two essential facts like Fact1 and Fact2 as secondary results.

Fact1: The Relation of NM and LDF-Coefficient

Until now, no researchers found the relation of NM and LDF-coefficient. The pattern recognition researchers explain the relationship between NM and coefficient.

However, they could not explain the clear relation because the intercept is a free variable. IP-OLDF can explain the relation of NMs and discriminant coefficients clearly (Fact1). Let us consider the discrimination of three cases and two variables ($n = 3$, $p = 2$) as follows:

$$\begin{aligned} &\text{Class1}(y_1 = 1) \quad : \mathbf{x}_1 = (-1/18, -1/12). \\ &\text{Class2}(y_2 = y_3 = -1) \quad : \mathbf{x}_2 = (-1, 1/2), \mathbf{x}_3 = (1/9, -1/3). \end{aligned}$$

Equation (3) expresses IP-OLDF. To multiply y_2 and y_3 changes the signs of $\mathbf{x}_2$, $\mathbf{x}_3$ and constant. The role of y_i aligns the inequality signs with $\mathbf{x}_1$.

$$\begin{aligned} &\text{MIN} = \sum e_i; \\ &y_1 * \{-(1/18) * b_1 - (1/12) * b_2 + 1\} \geq -e_1; \\ &y_2 * \{-b_1 + (1/2) * b_2 + 1\} \geq -e_2; \\ &y_3 * \{(1/9) * b_1 - (1/3) * b_2 + 1\} \geq -e_3; \end{aligned} \tag{3}$$

We consider three linear equation (or discriminant hyperplane) in (4).

$$\begin{aligned} &\text{H1} = -(1/18) * b_1 - (1/12) * b_2 + 1 = 0, \\ &\text{H2} = b_1 - (1/2) * b_2 - 1 = 0, \\ &\text{H3} = -(1/9) * b_1 + (1/3) * b_2 - 1 = 0 \end{aligned} \tag{4}$$

Three linear hyperplanes divide the two-dimensional coefficient space into seven CP showed in Fig. 1. Each inner point has a different NM and misclassifies the same cases. IP-OLDF finds the vertex of correct OCP if data satisfy the general position. RIP finds the inner point of correct OCP and MNM directly. If other LDFs find the vertex or edge of CP and there are several cases on $f(\mathbf{x}) = 0$, correct NM may increase. Thus, NM is not reliable statistics.

Figure 1 shows the relation between NM and LDF. Each point corresponds to the coefficients of LDF in the discriminant coefficient space and the linear hyperplane in the p-dimensional data space. All inner points have the same NM. This truth means all LDFs belonging to the same CP discriminate the same cases correctly and misclassifies other cases. If LDF chooses the vertex or edge of CP having over p-cases on the discriminant hyperplane ($f(\mathbf{x}) = 0$), we cannot decide whether those cases belong to class1 or class2. However, many papers define all cases on $f(x) = 0$ belong to class1. This understanding is not logical. Most researchers do not understand the true discriminant rule. We could not determine the class of cases on $f(x) = 0$. Because RIP finds the inner points directly and avoids to choose the vertex or edge, only RIP can solve Problem1. Other LDFs cannot find the correct NM.

Fact2: MNM Monotonic Decrease ($MNM_k \geq MNM_{(K+1)}$)

Let us MNM_k be MNM of k-independent variables. $MNM_{(k+1)}$ is MNM of (k + 1)-independent variables to add one variable to the former model. MNM_k is greater than and equal to $MNM_{(k+1)}$ because of CP_k is included in $CP_{(k+1)}$. If $MNM_k = 0$, all models including these k-variables are LSD because of MNM monotonic decrease. This Fact2 indicates microarrays include an enormous number of the linearly separable gene subspaces such as Russian doll Matryoshka. When we discriminate Swiss banknote data with six variables [13], IP-OLDF finds two-variables such as (X4, X6) is LSD and the minimum dimensional SM (Basic Gene Set, BGS). By the MNN monotonic decrease (Fact2), 16 MNMs including BGS are zero among 63 models (= $2^6 - 1 = 63$). Other 47 MNMs are higher than one. This truth shows the Matryoshka structure of LSD. Method2 and RIP can find SMs from ordinary data and six microarrays naturally. Because we have developed the best model, we overlooked that RIP can select features naturally from ordinary data. Fisher validated Fisher's LDF by Fisher's iris data [29] that consists of three species (150 cases) and four independent variables. Although two species such as virginica and versicolor (100 cases) are not LSD, 15 models show many coefficients of IP-OLDF are naturally zero [23]. This fact indicates the following essential matters:

(1) Because MP-based LDF makes some coefficients zero for overlapping data, to make coefficients zero is irrelevant for LSD-discrimination.
(2) Thus, we do not expect that LASSO is useful for Problem5 because it cannot discriminate LSD theoretically.
(3) LSD is more valuable signal than to make coefficients zero.

The triangle in Fig. 1 is OCP (feasible region) and its MNM = 0. This truth means all points of OCP are the optimal solution. Because the purpose of MP is to find an optimal solution from within the feasible region, our OLDF is not a typical MP model. For LSD, three OLDFs and three SVMs share the same OCP (feasible region). This point is different from other researcher's SVMs those are defined in the whole space.

2.3 Valuable Basic Knowledge of Linear Algebra and MP

2.3.1 Basic Knowledge of Linear Algebra

The basic knowledge of linear algebra tells us about the solution of simultaneous equations. In an ordinary sample (n > p), the vertex of OCP is less than p-dimension. If we choose p linear equations from n linear equations, we can calculate the p-variables solution in the p-dimensional data. In high-dimensional microarray ($n \ll p$), the vertex is less than an n-variables solution in p dimensional data space. If we choose n-variables and ignore (p-n) variables, we can calculate the n-variables solution in the n-dimensional data. Therefore, we create (p * p) regular matrices by selecting p linear equations from n linear equations in the case of (n cases > p variables) or create (n * n) regular matrices by selecting n variables from p variables in the case of (n cases < p variables). Then, Revised LP-OLDF find a solution from these regular matrices.

2.3.2 Simplex Method

The algorithm of LP uses the simplex method that chooses the optimal vertex by calculating (p * p) matrix for ($n \geq p$) or (n * n) matrix ($n \ll p$). If we combine this knowledge with Fig. 1, we can understand the LINGO Program3 can find the optimal solution quickly by both RIP and Revised LP-OLDF. The first step of the simplex method finds the number of genes less than n. This result shows the number of non-zero coefficients less than n [31–33]. There are many OCPs because n linear hyperplanes divide microarray into many CPs those are less than n-dimensional subspaces. We currently do not know the meaning of the priority of the order in which Program 3 chooses SM. Also, although we consider that it is rare to select another vertex of the same OCP, it cannot be concluded that the possibility is zero. Also, both feasible regions of RIP and Revised LP-OLDF are the same CP, but we do not know RIP choose which point. It may be the center of gravity of the OCP.

2.4 *MP-Based LDFs*

2.4.1 Three Optimal LDFs

If data is not general position, IP-OLDF may not look for the correct vertex of OCP because there may be over (p + 1) cases on $f(\mathbf{x}) = 0$, and we cannot correctly discriminate these cases. In this case, the obtained NM is not correct and increases. Thus, we developed RIP that looked for the inner point of true OCP directly in (5). M is 10,000 (Big M constant). Because b_0 is a free variable, it is defined in (p + 1)-dimensional coefficient space as same as the pattern recognition. If it discriminates $\mathbf{x}_i$ correctly, $e_i = 0$ and $y_i*({}^t\mathbf{x}_i*\mathbf{b} + b_0) \geq 1$. If it cannot discriminate $\mathbf{x}_i$ correctly, $e_i = 1$ and $y_i*({}^t\mathbf{x}_i*\mathbf{b} + b_0) \geq -9999$. Thus, support vector (SV) for classified cases choose $y_i*({}^t\mathbf{x}_i*\mathbf{b} + b_0) = 1$ and SV for misclassified cases choose $y_i*({}^t\mathbf{x}_i*\mathbf{b} + b_0) = -9999$. Because DSs of misclassified cases are less than -1, there are no cases within two SVs. However, if M is a small constant, it does not work correctly [25]. Because no cases are on the discriminant hyperplane, we can understand the optimal solution is an inner point of OCP defined by IP-OLDF. All LDFs except for RIP cannot solve Problem1 theoretically. Thus, these LDFs must check the number of cases (k) on the discriminant hyperplane. Correct NM may increase (NM + k).

$$\mathrm{MIN} = \sum e_i;\ y_i * \left({}^t\mathbf{x}_i * \mathbf{b} + b_0\right) \geq 1 - M * e_i; \tag{5}$$

If e_i is non-negative real variable, Eq. (5) becomes Revised LP-OLDF. If it can discriminate some cases by SV correctly, those e_i become zero. Otherwise, e_i become positive real variable. It permits some cases to violate the distance "M*e_i" from SV. If we set M = 1, the objective function is to minimize the summation of distance by misclassified cases from SV. Because it may tend to collect several cases on the discriminant hyperplane (Problem1), we recommend not to use Revised LP-OLDF for the overlapping data. However, it can decompose microarrays into plural SMs as same as RIP. Revised IPLP-OLDF is a mixture model of Revised LP-OLDF and RIP. It can decompose microarrays, also. We ignore Revised IPLP-OLDF in this research. The optimal discriminant coefficient is a vertex of feasible region (CP) that simultaneously satisfies n linear inequality constraints. In summary, Revised LP-OLDF selects one vertex (endpoint) of this CP as an optimum solution. This point is the solution of n simultaneous equations by setting the arbitrary discriminant coefficient of (p − n) genes to 0. For LSD, RIP has the same OCP and the integer variable e_i do not affect the model because of $e_i = 0$. However, we cannot explain which point it chooses now. However, RIP and Revised LP-OLDF can easily avoid three difficulties. On the other hand, the algorithm of IP solver uses the branch & bound (B & B) algorithm that is as same as to search all subspaces of microarray efficiently. This function may affect to find Fact4.

2.4.2 Hard-Margin SVM and Soft-Margin SVM

Equation (6) defines H-SVM. Before H-SVM, nobody can define whether data is LSD or overlap. Many researchers believe that LSD-discrimination is easy. Now, LSD is defined by "MNM = 0," and overlap data is "MNM ≥ 1". However, H-SVM causes the computation error for the overlap data. This fact may be the reason why many researchers did not use H-SVM to discriminate microarrays. Moreover, because QP defines SVM and can find only the minimum solution of the whole domain, SVM cannot decompose microarrays into many SMs.

$$\mathrm{MIN} = ||b||^2/2;\ y_i * \left({}^t\mathbf{x}_i * \mathbf{b} + b_0\right) \geq 1; \tag{6}$$

e_i: non-negative real value.

Because most real data are not LSD and H-SVM causes a computational error for LSD, Vapnik defines soft-margin SVM (S-SVM) using penalty c in (7). S-SVM permits certain cases that are not discriminated by SV ($y_i \times ({}^t\mathbf{x}_i\mathbf{b} + b_0) < 1$). The second objective is to minimize the summation of distances of misclassified cases (Σe_i) from SV. The penalty c combines two objects and makes one objective function. Nevertheless, Revised LP-OLDF minimize the summation of misclassified distance from the discriminant hyperplane as same as the second objective function in (7). This fact shows two defects of SVMs as follows:

(1) Although Revised LP-OLDF can select SM, H-SVM and S-SVM cannot select SM because of the first objective function are the quadratic function.

(2) S-SVM does not have the rule to determine a proper c as same as RDA; nevertheless, an optimization solver solves SVMs.

Thus, we compare two S-SVMs, such as SVM4 (c = 10000) and SVM1 (c = 1). In many trials, NMs of SVM4 are less than NMs of SVM1.

We claim the methods with tuning parameters such as S-SVM and RDA are useless for standard statistical users because they must pay their efforts to select the best parameters for each data. On the other hand, although RIP must set the big M constant, we confirmed M = 10000 causes good results using six different types of common data and all possible models. We surveyed and investigated to change the value M from c = 0.1, 1, 10, 100, 10^3, 10^4 and 10^6 [25].

$$\mathrm{MIN} = ||\mathbf{b}||^2/2 + c \times \sum e_i;\ y_i \times ({}^t\mathbf{x}_i\mathbf{b} + b_0) \geq 1 - M * e_i \quad (7)$$

c penalty c for combining two objectives.
e_i non-negative real value.
M big M constant.

2.4.3 Other MP-Based LDFs

Stam [38] summarized about 140 papers and books on MP-based discriminant functions and discussed in his paper the topic "Why statistical users rarely used MP-based discriminant functions (Lp-norm classifies)." Our opinion is simple as follows:

(1) Fisher firstly established discriminant analysis as a statistical method. MP-based discriminant functions are follow-up studies of statistical discriminant analysis. In this case, the latter need to show advantages over statistical methods by empirical studies. However, many studies defined only the models and ended in 1999 after Stam paper. In contrast, we compared RIP with the other seven LDFs by Method1. Six types of data showed the following rank:

  (a) RIP is almost the best rank,
  (b) logistic regression, SVM4, Revised LP-OLDF and Revised IPLP-OLDF are the second ranks,
  (c) Fisher's LDF and SVM1 are the worst

(2) RIP based on MNM criterion defines OCP as an executable region. H-SVM maximizes the distance between SVs. Only these LDFs could demonstrate superiority in empirical studies to statistical discriminant functions. In other words, the researches of the first generation MP-based discriminant analysis were ended by Stam's review paper in 1997. In the future, we must study the reason for the failure of these models.

3 First Success of Cancer Gene Analysis

Because statistical discriminant functions are useless, medical researchers could not complete their researches. Other researchers approached the cancer gene analysis as new topics from the viewpoints of statistics or engineering. They pointed out three difficulties of the cancer gene analysis and could not succeed in it as same as the medical researchers (Problem5). However, we firstly succeed in Problem5 using six microarrays.

3.1 Three Difficulties of Problem5

Other researchers pointed out three difficulties of Problem5 as follows:

3.1.1 Small n and Large p Data [10]

To estimate the variance-covariance matrices of microarrays was difficult for statistical discriminant functions. On the other hand, MP-based LDFs are free from this problem. Sall [21] announced Fisher's LDF for high-dimensional microarrays by the singular value decomposition (SVD) at the Discovery Summit in Tokyo, November 10, 2015. When we borrowed a new version of JMP and discriminated microarrays by Fisher's LDF, six NMs (error rates) were large. This fact shows the defect of discriminant functions based on variance-covariance matrices. Also, Fisher's LDF by maximizing the correlation ratio could not discriminate LSD correctly for many data. Only two standards such as the maximization of SV distance by H-SVM and MNM criterion of RIP can discriminate LSD theoretically. Moreover, six MP based LDFs discriminate small n and large p data compared with large n and small p data quickly.

3.1.2 Feature Selection Is NP-Hard [6]

General speaking, it is difficult to select proper cancer genes for large p-genes. Many statisticians do not understand that we must find one of the optimal solutions (MNM = 0) of subspaces in p-dimensional space. In general, the statistical discriminant function finds one LDF on the p-dimensional domain. To find the optimum LDF with MNM = 0 of the subspace needs all possible model [16] to search all subspaces as same as B&B algorithm. If p is large, it will surely be NP-hard. However, RIP and Revised LP-OLDF can find one of the optimum LDFs of subspaces. Because QP finds the only optimal SVM to maximize/minimize the objective function in the p-dimensional space, it needs feature selection and is NP-hard.

3.1.3 Signal Subspace Is Embedded in Noise Subspace [4]

This difficulty is unclear because the definitions of signal and noise are not clear. Until now, they could not specify the definition of the signal except for Golub et al. (signal and noise statistics). Because all SMs are LSD, we consider each SM is signal because two-classes are separable in SM. Although to measure microarrays is expensive, to measure SM saves the expense for cancer gene diagnosis if we can decide the best SM for cancer gene diagnosis.

3.2 Cancer Gene Analysis by Six Microarrays

Jeffery et al. [17] uploaded six microarrays of Table 1. They compared the efficiency of the ten feature selection methods. Although their researches summarize the established feature selection methods, those methods could not find Fact3. When RIP discriminated microarrays, we found those are LSD (Fact3). Because no researchers knew this fact, they could not succeed in Problem5. LSD was the most important signal.

When RIP discriminated Shipp data, we found only 32 coefficients are not zero. After many trials, we found the Matryoshka feature selection method by Golub data. Everyone can download 15 free papers about the detail process of Method2 from Research Gate (https://www.researchgate.net/profile/Shuichi_Shinmura). In the first step1, RIP discriminate 7,129 genes, and 72 coefficients are not zero. After we discriminate 72 genes in step2, 46 coefficients are not zero. In step3, 36 coefficients are not zero. However, because we cannot reduce 36 genes again, we call it the first SM1. After we removed these 36 genes from 7,129 genes and made the second big Matryoshka, we discriminated 7,093 genes and obtained SM2. Thus, we made LINGO Program3 (Method2) of six MP-based LDFs. However, we find that three SVMs could not decompose microarrays into SMs. Because QP finds only one optimal solution on the whole domain, it must survey all possible models to find SM. It is NP-hard as same as statistical discriminant functions.

Thus, LINGO Program3 can decompose microarrays into many SMs (Fact4) and release from the curse of high-dimensional data. Moreover, three difficulties become only three excuses. Most researchers do not notice the benefits of MP.

4 First Success of Cancer Gene Diagnosis

At first, we considered succeeding in cancer gene diagnosis by statistical methods because all SMs are signals. However, we could not obtain useful results (Problem6). Therefore, we judged the statistical analysis of gene data were not useful for cancer gene diagnosis. By the breakthrough of signal data, we can succeed in the cancer gene diagnosis because all results show the linearly separable facts.

4.1 Problem6

4.1.1 The Failure of Statistical Analysis of SMs (Problem6)

Statistical analysis of all SMs was a failure. One-way ANOVA, t-test, correlation analysis, cluster analysis, and PCA cannot show the linearly separable facts. Most NMs of Fisher's LDF and QDF are not zero. These defects are caused by the following reasons:

(1) Fisher's assumption based on the normal distribution is not adequate for the rating, pass/fail determination and medical diagnosis, especially microarrays.
(2) To construct variance-covariance matrices for large p is difficult.
(3) The maximization of the correlation ratio is irrelevant for LSD-discrimination.

4.1.2 Comparison of Logistic Regression and Other Methods

On the other hand, all NMs of SMs by logistic regression are zero. The maximum likelihood developed by Fisher obtain the logistic coefficients that can find the linearly separable fact based on the real data structure without the influence of normal distribution. This truth is the reason why it can discriminate all SMs empirically. These results raise the severe problem of why there is no study showing Fact3 in AI research which infers the logistic coefficient from big data. Cilia et al. [8] analyzed Alon and Golub microarray by four classification methods such as decision tree, random forest, nearest neighbor, and multilayer perceptron. They introduced the best results are as follows. Recognition rates (RR) of Alon and Golub are 91.94 and 99.44% using 10 and 51 genes. Two RRs of Fisher's LDF are 92 and 89% using 2,000 genes and 7,129 genes. Cilia's results are better than JMP's results. They found these results by less than 51 genes. However, all RRs of SMs are zero by RIP and H-SVM. We guess LASSO [5] and RDA are useless for LSD-discrimination.

4.2 Analysis of 64 SMs of Alon

4.2.1 RatioSVs of 64 SMs

Because we obtain almost the same results from six microarrays, we explain 64 Alon's SMs found by RIP in 2016. At first, we discriminate all SMs by logistic regression, Fisher's LDF and QDF in addition to six MP-based LDFs. RIP and H-SVM discriminate LSD theoretically. Revised LP-OLDF, Revised IPLP-OLDF, SVM4, and logistic regression can discriminate all SMs empirically. However, SVM1, Fisher's LDF and QDF cannot discriminate all SMs correctly. Table 2 shows this result that is sorted by RatioSV (= SV distance*100/RIP DS) in descending order. SN column shows the sequential number. Gene shows the number of genes included in each

Table 2 NMs OF 64 SMs (2016)

SN	Gene	Logistic	QDF	LDF2	DS	RatioSV
8	31	0	0	0	7.5	26.8
35	30	0	0	1	8.5	23.5
11	25	0	0	2	9.7	20.7
	....					
59	36	0	0	6	37.2	5.4
14	26	0	0	0	39.8	5
64	42	0	0	8	84.9	2.4

Table 3 ONE-WAY ANOVA OF SM8

Gene	Min	Max	MIN	MAX	t (≠)	t (=)
X1473	4.74	8.12	4.89	9.55	4.20	4.27
X698	6.37	8.53	6.50	10.46	3.95	3.43
X1896	3.91	6.60	4.17	6.75	2.27	2.51
	....					
X14	11.18	12.84	10.18	12.84	−5.24	−5.12
X1423	6.03	10.75	3.50	9.84	−6.68	−6.50

SM. Logistic, QDF and LDF2 show NMs. DS is the range of discriminant scores of RIP. RatioSV equals to 200/DS. Because two classes of Alon data are well separable than other five microarrays, QDF can discriminate 64 SMs correctly in addition to logistic regression. The range of RatioSV is [2.4, 26.85%]. The 63 RatioSV is over 5%. These results show that SV separates two classes of 63 SMs over 5% width.

4.2.2 Analysis of the Eighth SM8

Because all SMs are small samples and LSD, we expect statistical methods find linearly separable facts. However, we cannot obtain useful results. SM8 contains 31 genes and takes the maximum RatioSV 26.8%. In Table 3, "Min and Max" and "MIN and MAX" show the range of two classes and show all genes overlap. The ranges of t-values spread from negative to positive values. This truth indicates that genes included in SM are not signal and useless to find oncogenes. Although many medical researchers analyze one or several genes by t-test, correlation, cluster analysis, they may not find useful results (Problem6).

Figure 2 is the output by PCA. The left plot shows the eigenvalues. The first eigenvalue is 4.6378, and the contribution ratio is 15%. Because the second eigenvalue is 3.8879 and the contribution ratio is 12.55%, the cumulative contribution ratio is 27.5%. Usually, we consider that both principal components can catch useful information because those represent about 30% data variation. The middle scatters plot shows two classes overlap. Right factor loading plot shows the correlation of 31 genes with the first principal component (Prin1) and the second principal com-

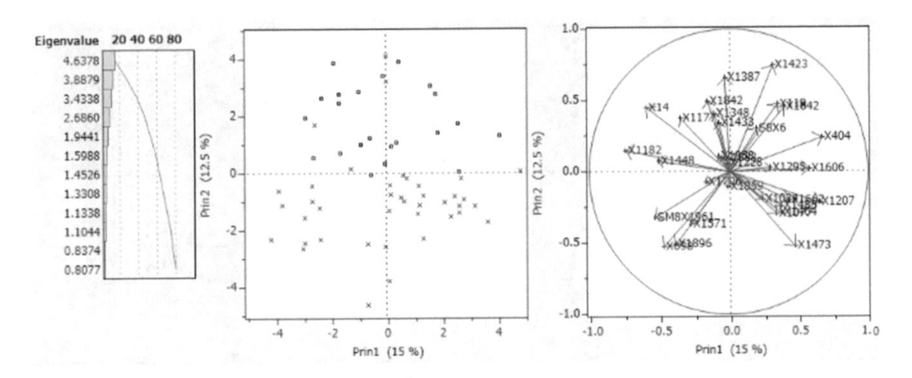

Fig. 2 PCA plots (eigenvalue, scatter plots, and factor loading)

ponent. We cannot find linearly separable facts from scatters plot and factor loading plot. Other statistical methods cannot explain the useful meaning of genes, also.

4.2.3 Breakthrough by RIP Discriminant Scores (RipDSs)

Table 4 shows the RIP discriminant score (RipDS), RatioSV and t-value. Because we set two SVs such as SV = −1 and 1, the SV distance is 2. The definition of RatioSV is (2 * 100/the range of DS). At first, we misunderstand the genes of SM as oncogenes. However, statistical methods did not find the linearly separable fact. Only RIP and H-SVM can discriminate LSD theoretically. After trial and error, we understand that RipDSs, LpDSs, and HsvmDSs are the actual signals of the microarray. Through this breakthrough, we make the signal data that consists of 62 cases and 64 RipDSs. We can reduce Alon microarray (62 cases and 2,000 genes) into signal data (62 cases and 64 RipDSs).

Table 4 The range of 64 RipDS and RatioSV (=200/DS)

RipDS	Min	Max	MIN	MAX	RatioSV	t (≠)
RipDS8	−3.35	−1	1	4.12	26.76	15.50
RipDS35	−2.58	−1	1	5.92	23.52	13.02
RipDS11	−4.15	−1	1	5.52	20.68	12.71
	…….					
RipDS59	−15.25	−1	1	21.91	5.38	6.17
RipDS14	−21.94	−1	1	17.85	5.03	7.16
RipDS64	−3.94	−1	1	81.00	2.35	4.22

4.3 Cancer Gene Diagnosis

We introduce the results of cancer gene diagnosis only by cluster analysis and PCA. Alon and Singh consist of cancer and healthy classes. Other four microarrays consist of two different types of cancers. Moreover, cancers are different. However, all results are almost the same. Thus, we claim that microarrays are useful for cancer gene analysis and diagnosis.

4.3.1 Ward Cluster Analysis

Figure 3 shows the cluster analysis of signal data that consists of 62 cases and 64 RIP DSs. The upper 22 normal subjects and lower 40 cancer patients become two clear clusters. Right side dendrogram shows the case dendrogram. The normal subjects become one cluster. The cancer cases consist of four clusters. Because the bottom two clusters consist of two patients, those may be the outlier and become the new subclasses of cancer pointed by Golub et al. The bottom dendrogram shows the variable clusters of 64 RipDS. The third and fourth RipDSs from the left first become clusters. Such pair correlation is often 1, and it is considered that different gene pairs are compatible and complementary to each other. That is, it may be an SM important for diagnosis.

4.3.2 PCA

Figure 4 shows the output of the PCA. Healthy subjects locate on the negative axis of Prin1 and cancer patients spread in first and fourth quadrants as they move away from the origin. However, because the eigenvalue of Prin1 in the left plot is 61.5% of the whole and Prin2 is 2.55%, the variation of Prin2 is tiny. Because the middle scatters plot shows that two groups are almost on the axis of Prin1, that represents another malignancy index in addition to 64 RipDSs. Aosima and Yata [2, 3] found that two classes locate on two different spheres in six microarrays. They found this fact by the high-dimensional PCA. Thus, we claim other microarrays may be LSD if data is well controlled. Furthermore, we find two clusters of two cases dendrogram locate on the first and fourth quadrants. Right factor loading plot shows the correlation between 64 RipDSs and two principal components. The 48th, 55th and 56th RipDSs are outliers. When we analyze the transposed signal data, we find many outliers. We expect those are new subclasses of cancer pointed out by Golub et al. Correlation analysis of signal data offers several facts. All correlations are positive values. Several correlations are 1. Those show that gene sets can be substituted for each other and redundant. The middle plot shows the score plot.

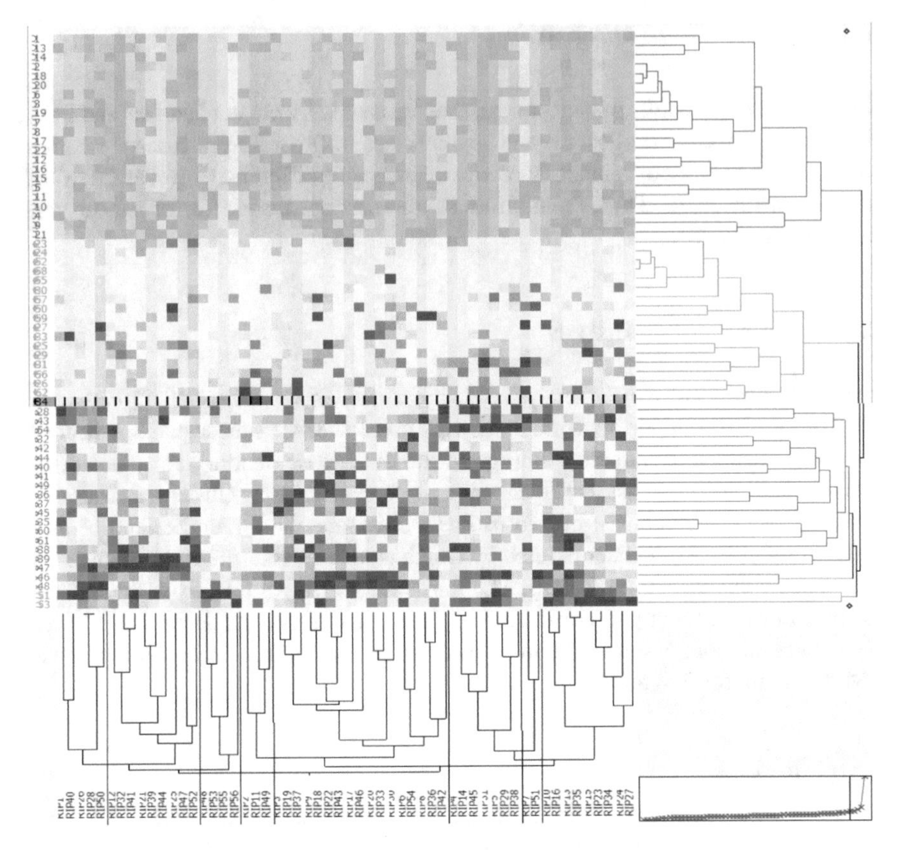

Fig. 3 The ward cluster of alon signal data (62 cases and 64 RIP DSs)

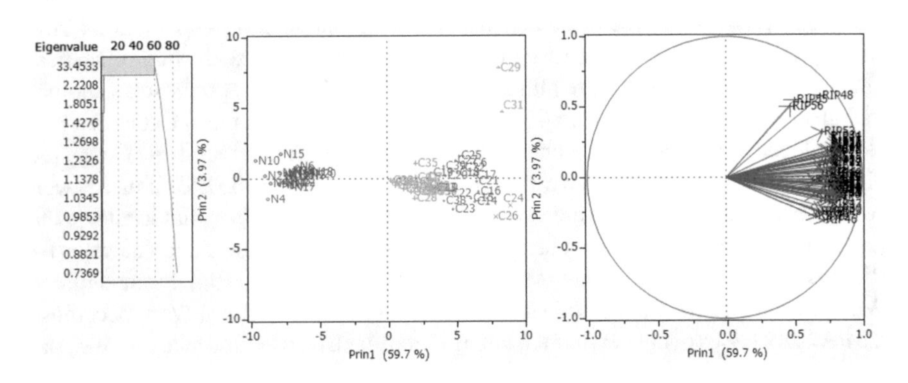

Fig. 4 PCA output (left: eigenvalue, center: score plot, right: factor loading plot)

5 Conclusion

Golub confessed they started their research around 1970. Because statistical discriminant functions were useless, they developed many original methods such as target marker genes, and weighted voting. Mainly, we guess their signal and noise statistics affected the third difficulty. Other medical projects often used Self-organizing map (SOM). One research uses the nearest neighbor clustering. Statisticians started this new theme as a high-dimensional data analysis. Engineers such as biology, bioinformatics, pattern recognition, and machine learning approached this theme by feature selection or filtering using many combinations of methods. However, no researchers solved Problem5 completely.

On the other hand, we solved Problem5 entirely within 54 days in 2015 as the applied problem of the new theory of discriminant analysis. Because the MP-based LDFs have solved four defects in the discriminant analysis, we can firstly succeed in the cancer gene analysis. That is, RIP quickly found that microarrays are MNM = 0 (Fact3). This fact shows that cancer and healthy subjects are LSDs in high dimensional data. Moreover, using the ordinary data we established the theory of LSD-discrimination, but Problem5 was just solved as applied research of this theory. Furthermore, LSD has a Matryoshka structure and includes many SMs (Fact4). We summarize our research as follows:

(1) RIP and H-SVM discriminate six microarrays and all SMs theoretically. Only RIP and Revised LP-OLDF can decompose many SMs and noise subspace. However, H-SVM could not find Fact4.
(2) Revised LP-OLDF and Revised IPLP-OLDF discriminate six microarrays and all SMs empirically.
(3) Logistic regression discriminates all SMs empirically because the maximum likelihood estimates the logistic coefficients.
(4) SVM4 and QDF often discriminate many SMs empirically.
(5) However, cluster analysis, PCA, one-way ANOVA and t-tests cannot show the linearly separable facts of all SMs. This matter is Problem6.
(6) Thus, although statistical discriminant functions are useless for Problem5, those are better than other statistical methods. This fact indicates us that large data variation by Prin1 and Prin2 cannot find linearly separable facts.

The only RIP and H-SVM can discriminate LSD correctly. Because the maximum likelihood method developed by Fisher solves the logistic regression, it can find real data structure without the influence of normal distribution. This truth is the reason why it can discriminate all SMs empirically. Therefore, we make signal data made by RipDSs, LpDSs, and HsvmDSs. By this breakthrough, we can offer the cancer gene diagnosis by statistical methods. However, after medical specialist validate and confirm our results of the cancer gene diagnosis.

LINGO [22] solved Problem5 and produced many SMs. RIP, Revised LP-OLDF, and H-SVM could make three signal data and solve Problem6. JMP [21] proposes the cancer gene diagnosis using signal data. Our claim is "good software enhances our intellectual ability." Without the cooperation of MP and statistics, we cannot solve Problem5 and Problem6 [34]. This is the reason why no researchers could find solve Fact3 and Fact4 of this research theme until 1970.

References

1. Alon, U., et al.: Broad patterns of gene expression revealed by clustering analysis of cancer and normal colon tissues probed by oligonucleotide arrays. Proc. Natl. Acad. Sci. USA **96**, 6745–6750 (1999)
2. Aoshima, M., Yata, K.: Distance-based classifier by data transformation for high-dimension, strongly spiked eigenvalue models. Ann. Inst. Stat. Math. **71**, 473–503 (2019)
3. Aoshima, M., Yata, K.: High-dimensional quadratic classifiers in non-sparse settings. Methodol. Comput. Appl. Probabil. (in press, 2019)
4. Brahim, A.B., Lima, M: Hybrid instance based feature selection algorithms for cancer diagnosis. Pattern Recognition Letters, pp. 8. 2014
5. Buhlmann, P., Geer, A.B.: Statistics for high-dimensional data-method, theory, and applications. Springer, Berlin (2011)
6. Charikar, M., Guruswami, V., Kumar, R., Rajagopalan, S., Sahai, A.: Combinatorial feature selection problems. IEEE Xplore, pp. 631–640 (2000)
7. Chiaretti, S. et al.: Gene Expression Profile of Adult T-cell Acute Lymphocytic Leukemia Identifies Distinct Subsets of Patients with Different Response to Therapy And Survival. Blood. April 1, 2004, 103/7, pp. 2771–2778 (2004)
8. Cilia, N.D., Claudio, D.S., Francesco, F., Stefano, R., Alessandra, S.F.: An experimental comparison of feature-selection and classification methods for microarray datasets. Information **10**(109), 1–13 (2019)
9. Cox, D.R.: The regression analysis of binary sequences (with discussion). J. Roy Stat. Soc. B 20 215–242 (1958)
10. Diao, G., Vidyashankar, A.N.: Assessing genome-wide statistical significance for large p small n problems. Genetics **194**, 781–783 (2013)
11. Firth, D.: "Bias reduction of maximum likelihood estimates. Biometrika **80**, 27–39 (1993)
12. Fisher, R.A.: Statistical Methods and Statistical Inference. Hafner Publishing Co., New Zealand (1956)
13. Flury, B., Riedwyl, H.: Multivariate Statistics: A Practical Approach. Cambridge University Press, New York (1988)
14. Friedman, J.H.: Regularized discriminant analysis. J. Am. Stat. Assoc. **84**(405), 165–175 (1989)
15. Golub, T.R. et al.: Molecular classification of cancer: class discovery and class prediction by gene expression monitoring. Science. 1999 Oct 15, 286/5439, 531–537 (1999)
16. Goodnight, J.H.: SAS Technical Report – The Sweep Operator: Its Importance in Statistical Computing – R (100). SAS Institute Inc. USA (1978)
17. Jeffery, I.B., Higgins, D.G., Culhane, C: Comparison and evaluation of methods for generating differentially expressed gene lists from microarray data. BMC Bioinformat. (2006)
18. Lachenbruch, P.A., Mickey, M.R.: Estimation of error rates in discriminant analysis. Technometrics **10**(1), 11 (1968)
19. Miyake, A., Shinmura, S.: Error rate of linear discriminant function. In: Dombal, F.T., Gremy, F. (ed.) North-Holland Publishing Company. The Netherland, pp. 435–445 (1976)
20. Miyake, A., Shinmura, S.: An algorithm for the optimal linear discriminant function and its application. Jpn Soc. Med. Electron Bio. Eng. **1815**, 452–454 (1980)
21. Sall, J.P., Creighton, L., Lehman, A.: JMP Start Statistics, Third Edition. SAS Institute Inc. 2004. (S. Shinmura, Supervise Japanese Version)
22. Schrage, L.: Optimization Modeling with LINGO. LINDO Systems Inc. (2006)

23. Shinmura, S.: Optimal Linear Discriminant Functions Using Mathematical Programming. Dissertation, Okayama University, Japan, pp. 1–101 (2000)
24. Shinmura, S.: A new algorithm of the linear discriminant function using integer programming. New Trends Probab. Stat. **5**, 133–142 (2000)
25. S. Shinmura, The optimal linear discriminant function, Union of Japanese Scientist and Engineer Publishing, Japan (ISBN 978-4-8171-9364-3), 2010
26. Shinmura, S.: Problem of discriminant analysis by mark sense test data. Japanese Soc. Appl. Stat. **4012**, 157–172 (2011)
27. Shinmura, S.: End of Discriminant Functions based on Variance-Covariance Matrices. ICORES, pp. 5–16 (2014)
28. Shinmura, S.: Four Serious Problems and New Facts of the Discriminant Analysis. In: Pinson, E., et al. (eds.) Operations Research and Enterprise Systems, pp. 15–30. Springer, Berlin (2015)
29. Shinmura, S.: New Theory of Discriminant Analysis after R. Springer, Fisher (2016)
30. Shinmura, S.: Cancer Gene Analysis to Cancer Gene Diagnosis, Amazon (2017)
31. Shinmura, S.: Cancer Gene Analysis by Singh et al. Microarray Data. ISI2017, pp. 1–6 (2017)
32. Shinmura, S.: Cancer Gene Analysis of Microarray Data. BCD18, pp. 1–6 (2018)
33. Shinmura, S.: First Success of Cancer Gene Analysis by Microarrays, pp. 1–7. Biocomp'18 (2018)
34. Shinmura, S.: High-Dimensional Microarray Data Analysis. Springer (2019)
35. Shinmura, S.: High-dimensional microarray data analysis—first success of cancer gene analysis and cancer gene diagnosis. August ISI2019, in Press (2019)
36. Shipp, M.A., et al.: Diffuse large B-cell lymphoma outcome prediction by gene-expression profiling and supervised machine learning. Nat. Med. **8**, 68–74 (2002)
37. Singh, D., et al.: Gene expression correlates of clinical prostate cancer behavior. Cancer Cell, **1**, 203–209
38. Stam, A.: Non-traditional approaches to statistical classifications: some perspectives on Lp-norm methods. Ann. Oper. Res. **74**, 1–36 (1997)
39. Tian, E., et al.: The role of the Wnt-signaling antagonist DKK1 in the development of osteolytic lesions in multiple myeloma. New Eng. J. Med. **349**(26), 2483–2494 (2003)
40. Vapnik, V.: The Nature of Statistical Learning Theory.Springer. 1999

Evaluation of Inertial Sensor Configurations for Wearable Gait Analysis

Hongyu Zhao, Zhelong Wang, Sen Qiu, Jie Li, Fengshan Gao and Jianjun Wang

Abstract Gait analysis has potential use in various applications, such as health care, clinical rehabilitation, sport training, and pedestrian navigation. This paper addresses the problem of detecting gait events based on inertial sensors and body sensor networks (BSNs). Different methods have been presented for gait detection in the literature. Generally, straightforward rule-based methods involve a set of detection rules and associated thresholds, which are empirically predetermined and relatively brittle; whereas adaptive machine learning-based methods require a time-consuming training process and an amount of well-labeled data. This paper aims to investigate the effect of type, number and location of inertial sensors on gait detection, so as to offer some suggestions for optimal sensor configuration. Target gait events are detected using a hybrid adaptive method that combines a hidden Markov model (HMM) and a neural network (NN). Detection performance is evaluated with multi-subject gait data that are collected using foot-mounted inertial sensors. Experimental results show that angular rate hold the most reliable information for gait recognition during forward walking on level ground.

Keywords Gait detection · Machine learning (ML) · Neural network (NN) · Hidden Markov model (HMM) · Body sensor networks (BSNs)

H. Zhao (✉) · Z. Wang · S. Qiu · J. Li
School of Control Science and Engineering, Dalian University of Technology, Dalian 116024, China
e-mail: zhaohy@dlut.edu.cn

F. Gao
Department of Physical Education, Dalian University of Technology, Dalian 116024, China
e-mail: swimclub@dlut.edu.cn

J. Wang
Beijing Institute of Spacecraft System Engineering, Beijing 100094, China
e-mail: wjjxy1998@163.com

R. Lee (ed.), *Big Data, Cloud Computing, and Data Science Engineering*, Studies in Computational Intelligence 844,
https://doi.org/10.1007/978-3-030-24405-7_13

1 Introduction

Gait analysis is the analysis of various aspects of the patterns when we walk or run, which is an active research issue for applications of health care, clinical rehabilitation, sport training, pedestrian navigation, etc. [14, 19, 22, 24]. Normal gait is achieved when multiple body systems function properly and harmoniously, such as visual, vestibular, proprioceptive, musculoskeletal, cardiopulmonary, and nervous systems. Therefore, gait performance is considered to be an indicator and predictor of general physical health and functional status of individuals [13, 17]. The validity and reliability of gait analysis depend strongly on the measuring methods used.

This paper extend the work that we have done in [25] by looking at how sensor configuration affect the accuracy, efficiency and reliability of gait detection when walking forward on level ground. This issue is addressed by seeking answers to the following questions:

(1) What accuracy of gait detection can be achieved when employing foot-mounted inertial sensors during level walking?
(2) How do the detection accuracy and efficiency change with the change in type, number and placement of inertial sensors?
(3) If any particular inertial measurements have more profound effect on detection performance than the others?

The detection performance was evaluated using metrics of sensitivity, specificity, and computational cost. Although the system is constructed of dual foot-mounted inertial sensors for demonstration purpose, the method can be easily adapted to new sensors and new sensor placements for gait analysis.

2 Related Work

After decades of development, quantitative instruments for human gait analysis have become an important instrumental tool for revealing underlying pathologies that are manifested by gait abnormalities. However, gold standard instruments for gait analysis, such as optical motion capture systems and force plates, are commonly expensive and complex while needing expert operation and maintenance, and thereby not pervasive enough even in specialized centers and clinics. At present, gait analysis in most clinics and health centers is still mainly achieved by patient self-reporting and clinician observation. These subjective and qualitative methods are only suitable for preliminary gait examination. Although some severe gait abnormalities can be visually observed, subtle differences might be overlooked without quantitative measurements.

Fortunately, in some applications, accuracy is not the only or primary concern for gait analysis, and other relevant concerns include simplicity, accessibility, portability, etc. For example, it might be more meaningful to monitor gait function of

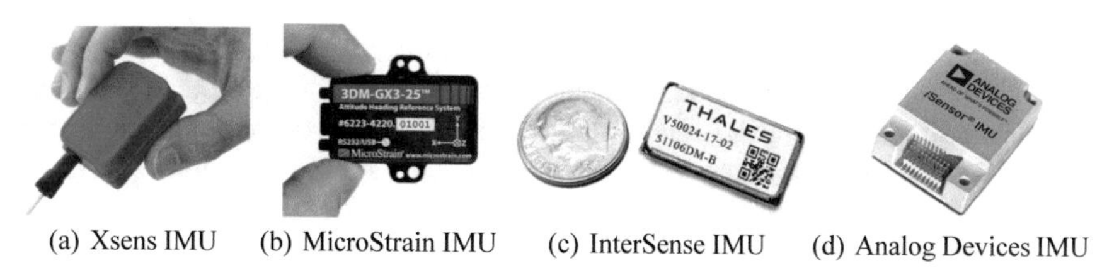

(a) Xsens IMU (b) MicroStrain IMU (c) InterSense IMU (d) Analog Devices IMU

Fig. 1 Typical MEMS inertial sensors

patients or elders in their living environments than just a brief gait examination in a laboratory or a clinic. Generally, subjects only exhibit natural gait patterns when they are accustomed to the walking environments; and clinical settings might affect their normal movement both mentally and physically, making the derived information fail to reflect the gait patterns in real-world settings. Thus, there is an increasing demand for a more accessible gait analysis system that can be used in both clinical trials and daily lives, so as to address the limitations of gold standard methods. Alternative instruments, such as inclinometers, goniometers, air pressure sensors, foot switches or force sensitive resistors, and inertial sensors, have been considered. These instruments are more portable, cost-effective, and easy-to-use, among which inertial sensors are widely considered attractive alternatives.

Due to recent developments in micro-electromechanical systems (MEMS), the cost, size, and weight of inertial sensors are decreasing whilst the accuracy is being improved, which provides an effective way for collecting human gait data with body sensor networks (BSNs). The commonly used MEMS inertial measurement unit (IMU) in the literature are shown in Fig. 1. For the pedal locomotion that is achieved by human leg motion, the intuitive experience is to perform gait analysis by attaching sensors to foot. As foot is the part of lower limb distal to leg, it acts as the interface between the lower limb and the ground, and hence has more obvious periodic nature than other parts of lower limb. This makes foot a preferred location of inertial sensors for gait data collection.

Undoubtedly, gait detection is a prerequisite for gait analysis. Researchers have presented different methods and sensor configurations for gait detection in the last few decades. The most commonly used detection methods could be divided into two categories, i.e., rule-based methods and machine learning-based methods. The rule-based detection methods are straightforward and easy-to-understand, but they involve a set of predefined rules and associated thresholds. At least three types of rules are involved in the detection process, i.e., peak detection, flat-zone detection, and zero-crossing detection. Besides, the involved thresholds are usually hand-tuned and fixed in the whole detection process regardless of gait changes. Meanwhile, the process of rule designing and threshold tuning itself is frustrating and time-consuming. Generally, the rule-based detection methods require a careful sensor alignment or parameter adjustment for each person's individual gait data, which are brittle or unable to adapt. As discussed in our previous study [23], for a good detection performance, the roles of all detection parameters need be carefully analyzed, as well as the relationship between them.

To make the potential use of gait analysis be fully exploited, it is necessary to develop an adaptive detection methods that can be easily adapted to new subjects and their new gait. Actually, gait detection is a pattern recognition problem, and hidden Markov models (HMMs) have been widely used for pattern recognition [3, 8, 11, 15]. HMM can model a Markov process with discrete and stochastic hidden states, which is in just one state at each time instant. For gait detection, the gait phases are the hidden states of HMM. Generally, pure HMMs work well for low dimensional data, but are less suitable for high dimensional data. Typically, a gait model can be driven by various combinations of direct or indirect inertial measurements, with the sensors attached to the subject's shank, thigh or lower lumbar spine near the body's center of mass (COM), etc. Thus, the obtained gait data might be of high dimension. Inspired by the existing methods, a neural network (NN) was adopted in our previous study to deal with the raw inertial measurements and feed the HMM with classifications [25]. This NN/HMM hybrid method takes advantage of both discriminative and generative models for gait detection, which can automatically capture the intrinsic feature patterns with no requirement of prior feature extraction and selection.

The contributions of this paper are as follows:

(1) Six gait events are involved for a detailed analysis of normal human gait, i.e., heel strike, foot flat, mid-stance, heel off, toe off, and mid-swing;
(2) Effect of type, number, and placement of inertial sensors on gait detection is investigated to offer some suggestions for sensor configuration when performing gait analysis.

3 Gait Data Acquisition

Gait analysis can be achieved by examining the inherent patterns of sensed gait data during waling or running. Two data sources are available for gait analysis in our study: inertial sensing system and optical motion capture system, as discussed in the following.

3.1 System Setup

In real-world settings, the hardware component includes two wearable nodes as data acquisition units and a portable device as data processing unit. Figure 2 shows the system setup and data acquisition process for gait analysis under normal conditions. In laboratory settings, optical motion capture system is used to provide the ground truth for training a gait model, which is the Vicon® system from Oxford Metrics Ltd., UK [2]. As shown in Fig. 3, a pair of inertial sensing nodes and three pairs of Vicon reflective markers are attached to the subject's feet.

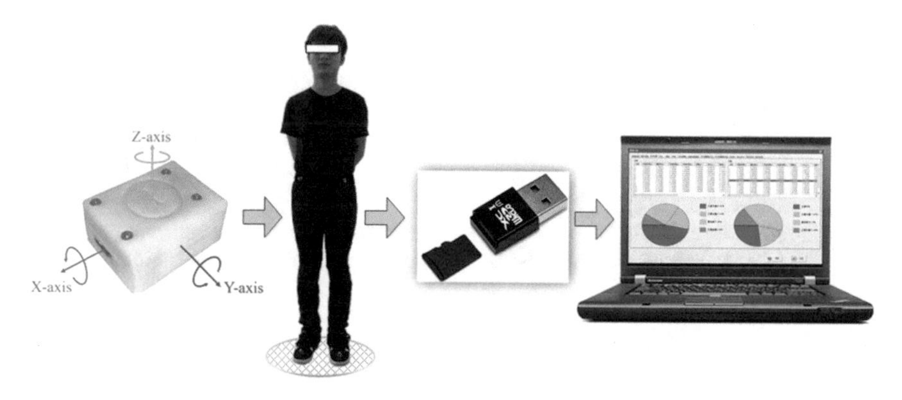

Fig. 2 Gait data acquisition in real-world settings

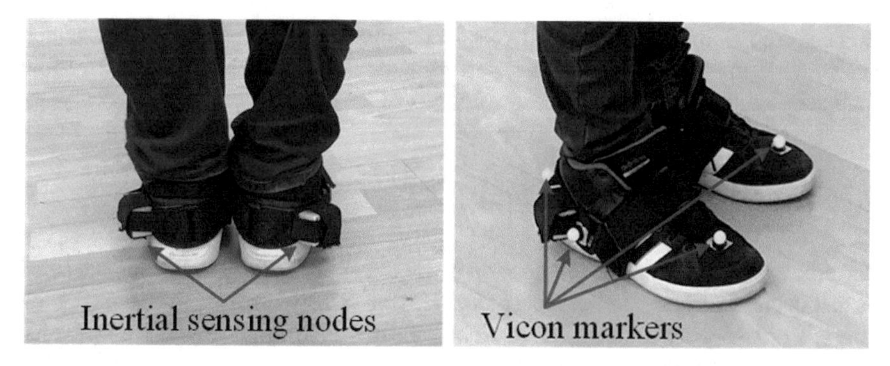

Fig. 3 Gait data acquisition in laboratory settings

3.2 *Inertial Sensor*

The IMU embedded in each sensing node is the ADIS16448 iSensor® device from Analog Devices Inc., USA [1], as shown in Fig. 4. Although each ADIS16448 has a triaxial gyroscope, a triaxial accelerometer, a triaxial magnetometer, and pressure sensors, only the measurements from the gyroscope and accelerometer are used for gait analysis in our study due to the indoor environmental conditions [7]. The main hardware components of each sensing node are the ADIS16448 IMU, a printed circuit assembly (PCA) with a microcontroller and its auxiliary circuits, a power supply, and a casing enclosing all the components. The dimensions of the entire sensing node are 4.5 cm × 3.5 cm × 2.25 cm, and the sampling rate is 400 Hz. Data collected during walking or running were stored in an internal memory first, and then transferred to an external computer for further processing.

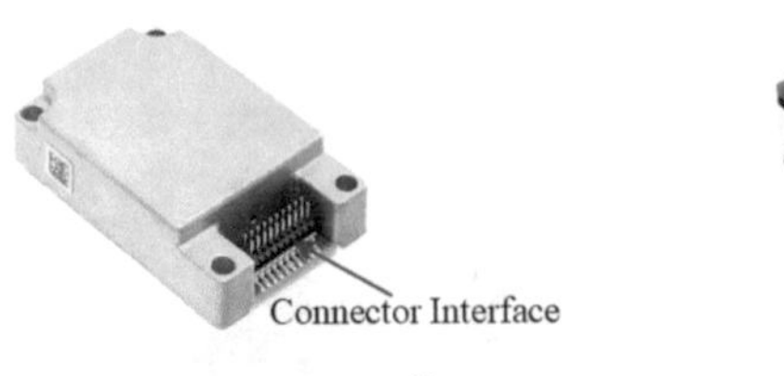

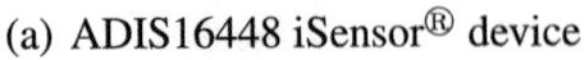
(a) ADIS16448 iSensor® device

(b) Sensing node without casing

Fig. 4 Hardware setup of foot-mounted inertial sensing nodes

4 Temporal Gait Parameter

To perform gait analysis, temporal gait parameters should be estimated first, which can provide measures to assess gait performance (such as symmetry, stability, and regularity) and delimit zero-velocity intervals to reset system errors periodically.

4.1 Gait Event and Gait Phase

Terminologically, gait is the locomotion mode that exhibits periodic patterns termed as gait cycle, and a stride refers to a gait cycle that starts with foot initial contact and consists of two consecutive steps [6]. For a normal gait, each cycle has a sequence of ordered gait events. Different researchers focus on different gait events according to their specific application. There are four typical events in one gait cycle, i.e., heel-strike (HS), foot-flat (FF), heel-off (HO), and toe-off (TO), as shown in Fig. 5 identified relative to right foot. These key events can divide a gait cycle into four consecutive time intervals termed as gait phases. Two additional events are taken into consideration in our study, i.e., mid-stance (MSt) and mid-swing (MSw), so as to give a detailed examination of human gait.

Examples of temporal gait parameters include, but are not limited to, cadence, stride (or step) duration, gait phase duration and percentage, the numbers of strides (or steps) taken, etc. These temporal parameters can be derived only if the gait events are correctly detected. Thus, gait detection is of particular relevance to gait analysis.

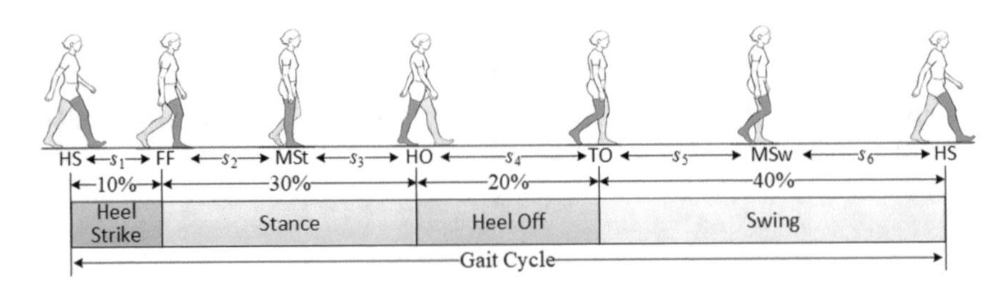

Fig. 5 Key events and phases in one gait cycle

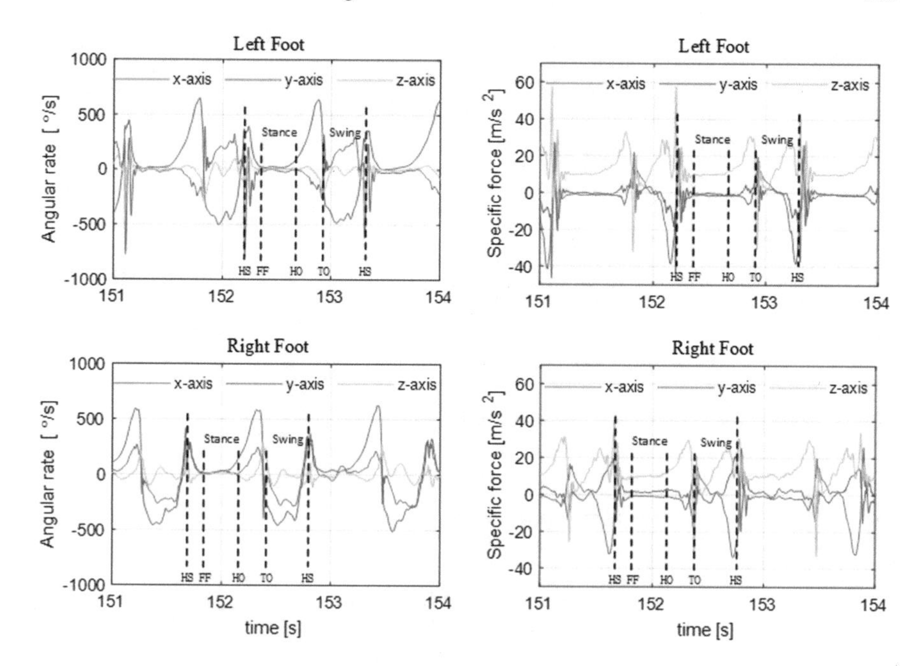

Fig. 6 Inertial measurements with key gait events and gait phases

4.2 Gait Data and Gait Division

A segment of raw inertial measurements (i.e., angular rate and specific force) is shown in Fig. 6, together with the identified gait events. As the IMUs are attached to feet, the measurements show periodic and repetitive patterns according to gait cycles. These patterns are helpful for gait analysis, by facilitating the detection of the key gait events and the concerned gait phases accordingly.

5 Gait Detection Method

In this section, commonly used methods for detecting gait events are first discussed, and then the implementation of an adaptive detection method is presented.

5.1 Rule-Based Detection Method

Generally, rule-based detection methods have two steps: first extract features with a sliding window technique from the inertial measurements, and then match the features with predefined rules and thresholds to identify distinct events in a sequence.

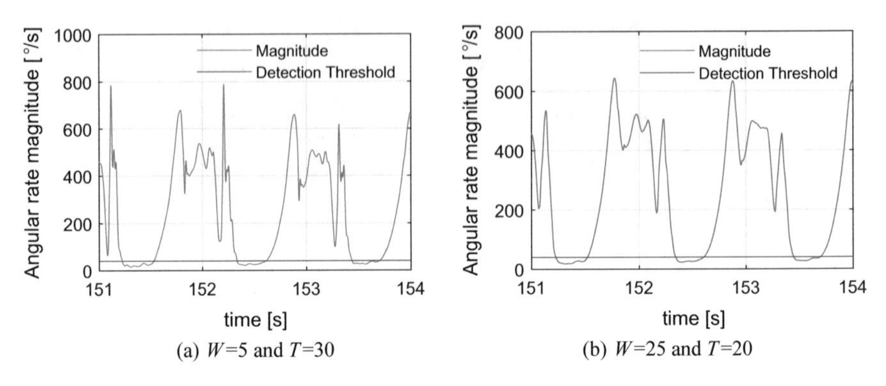

(a) W=5 and T=30

(b) W=25 and T=20

Fig. 7 Stance phase detection with angular rate magnitude

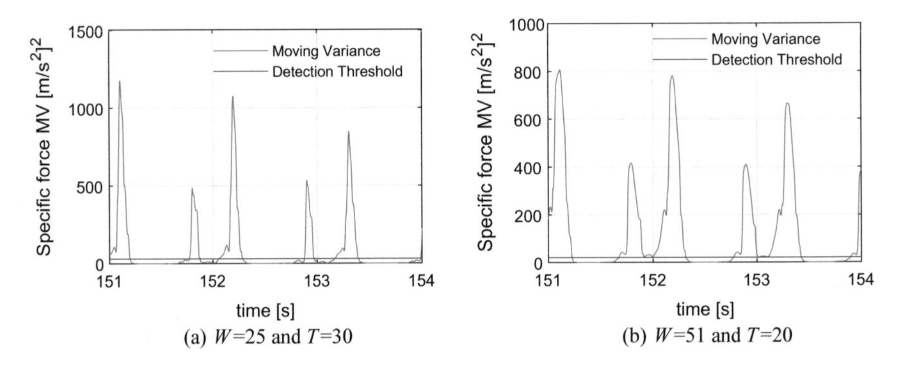

(a) W=25 and T=30

(b) W=51 and T=20

Fig. 8 Stance phase detection with acceleration moving variance (MV)

The extracted features can be magnitude [9, 16], root mean square [21], moving average [4, 5], moving variance [10], etc. For a dataset collected at 100 steps/min during normal walking, the flat-zone detection methods based on angular rate magnitude and acceleration moving variance are illustrated in Fig. 7 and Fig. 8 respectively, for the detection of merely stance phases that are delimited by FF and HO events, where W denotes the window size and T denotes the detection threshold. It can be seen that temporal fluctuations exist in the extracted features, which give rise to false gait phases when comparing to the detection threshold.

The above-mentioned rule-based detection methods have at least three improved variations, which are either more straightforward or more accurate, as discussed in the following.

- Method without contextual information: setting the window size to be one with no time delay, no signal distortion but fewer extractable features, e.g., measurement magnitude;
- Method without feature extraction: matching the raw measurements directly to the predefined rules and thresholds, e.g., the angular rate of pitch motion in the sagittal plane;

- Method with duration threshold: confirming the potential gait phases by looking at their durations using a time heuristic approach, as false phases are relatively short-lasting.

As discussed above, rule-based detection methods are generally brittle or unable to adapt. When handling new subjects, new motions, new sensors, and new sensor locations, new detection rules and associated thresholds are required. Thus, there is a pressing need for an adaptive detection method.

5.2 Machine Learning-Based Detection Method

An adaptive hybrid detection method is presented in our previous study [25], by modeling human gait with a six-state HMM and employing a three-layer NN to deal with the raw inertial measurements. This hybrid method is supposed to overcome the disadvantages of rule-based detection methods.

5.2.1 Left-to-Right HMM Model

During each sampling interval, the Markov process generates an observation, and then stays in the current state or transits to the next one, as shown in Fig. 9, where s_i denotes the hidden state, a_{ij} denotes the state transition probability, and $i, j \in [1, 2, \ldots, N]$. For the six-state gait model described in this paper, there is $N = 6$. This process yields a sequence of hidden states and a sequence of corresponding observations. Each state represents a gait phase that starts with the current gait event and continues until the next event. In normal gait, out-of-sequence events are not permitted due to the periodic nature of foot motion. Thus, each state can only transit to itself or its "right" state, and the HMM model has been trained using a left-to-right topology.

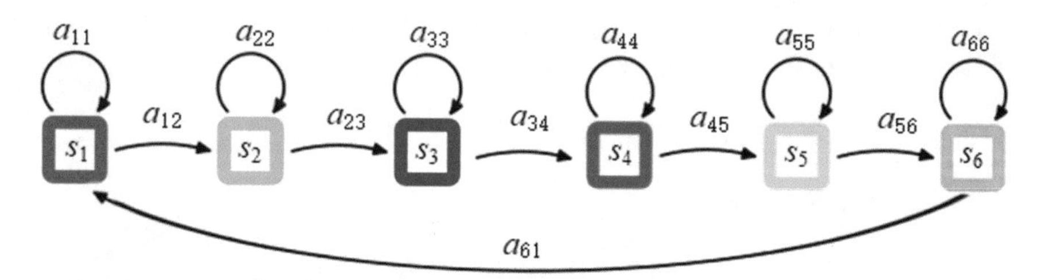

Fig. 9 Left-to-right HMM with six gait phases

5.2.2 Hybrid NN/HMM Model

Given a trained HMM and a sequence of time-ordered observations, the well-known Viterbi algorithm can estimate the most likely sequence of hidden states that have generated these observations. However, HMMs are generative models, compared to which discriminative models are supposed to achieve better classification performance. Discriminative models in machine learning, such as support vector machine (SVM) and k-nearest neighbor (k-NN), seem to be promising alternatives to HMMs for gait detection [18]. Previous studies suggest that NN allows the best trade-off between efficiency and accuracy. Theoretically, a three-layer network can approximate any nonlinear functions at any accuracy, given enough number of neurons in the hidden layer and enough training time [12]. However, NNs are limited to deal with each input element in isolation, rather than in context. To take advantage of both discriminative and generative models, an intuitive way is to effectively fuse them in the so-called hybrid manner.

The structure of the hybrid NN/HMM detection method is shown Fig. 10. The hybrid method has a training phase and a testing phase, which might be computationally complex in the training process, but computationally efficient at runtime. The NN is trained with the cross-entropy cost function and scaled conjugate gradient backpropagation algorithm. To train the HMM, the NN classifications are fed to HMM as observations, while the gait labels function as known HMM states. On the other hand, the HMM can model the sequential property of human gait, which complements the NN with contextual information. This hybrid method does not require a careful sensor alignment or parameter adjustment for each person's individual gait data, and generalizes well to new subjects, new motions, new sensors, and new sensor locations.

6 Experiment and Result

In this section, the hybrid NN/HMM method is utilized to implement gait detection first with varying type, number and placement of inertial sensors, and then some discussions are made on sensor configuration.

6.1 Experiment Setup

To evaluate the effect sensor configuration on gait detection, the experiments were carried out in a corridor of a typical office building, and nine healthy subjects were participated in the experiment. Each subject was asked to walk at self-selected speeds, and repeat the experiment four times along a 20 m straight and level trajectory. Prior to each trial, the subject stood still for a brief period to perform the initial alignment and calibration of the sensing system.

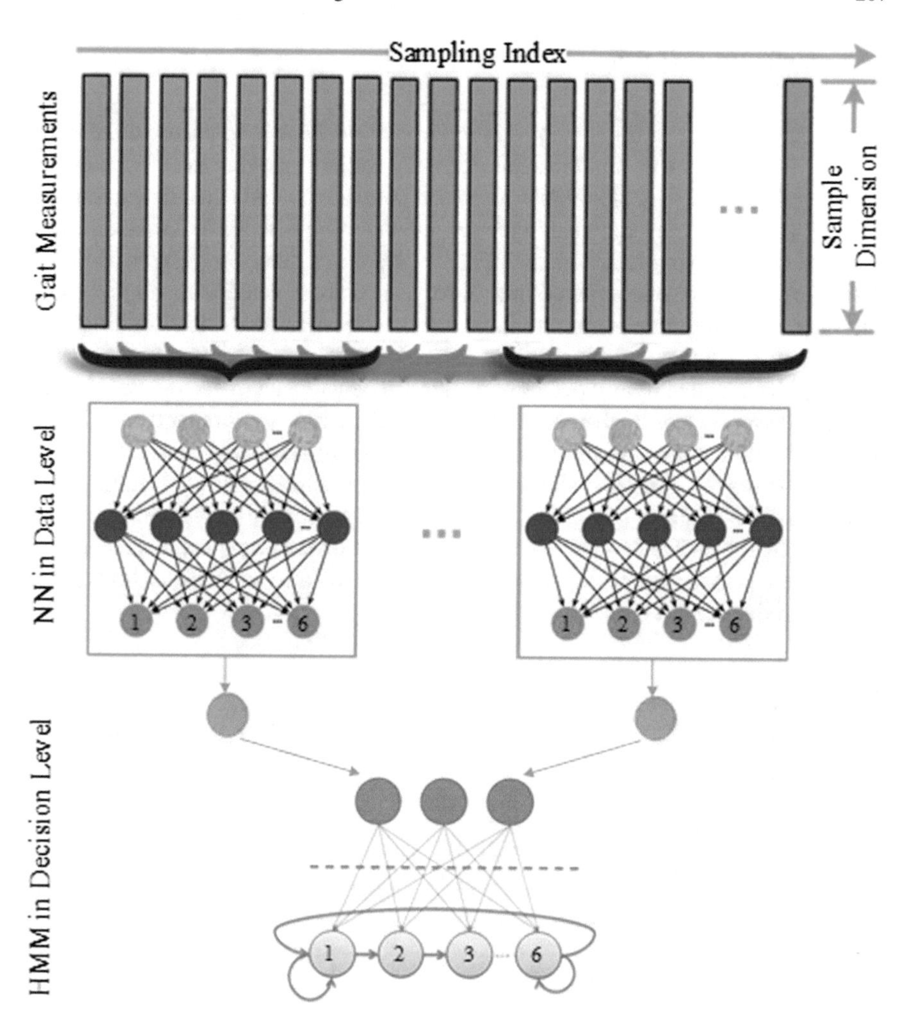

Fig. 10 Structure of the hybrid NN/HMM-based gait detection method

To train the hybrid gait model, each input of NN in the data level is formed from a sliding window of fifty-one samples with a step-size of one, the hidden layer consists of fifteen neurons, and the output layer consists of six neurons corresponding to the six gait phases. Here, a sample contains all interested gait data at the given time instant. The NN assigns a label to each input vector, which indicates the gait state at the central point of the sliding window.

6.2 Detection Result

The main challenge in implementing inertial gait analysis is to ensure a sufficient accuracy, especially to retain the same accuracy while improving the efficiency and reducing the complexity of both hardware and software components. In general, gait detection can be achieved by using the sensed data of gyroscope and accelerometer separately or by fusing them together. For the rule-based detection method, different data sources have been compared in [20] for stance phase detection, and the results suggest that angular rate is more reliable than acceleration for typical walking.

This paper compares different data sources for the detection of six gait phases, by using the adaptive hybrid detection method. The gait data in each sample shown in Fig. 10 can be from either uniaxial or triaxial inertial sensors, and from either unilateral or bilateral lower limbs. Eight sensor configurations are considered in this paper, and each configuration yields one kind of data source, as listed in Table 1. The detection performance is quantified using metrics of sensitivity and specificity, which are listed in Tables 2 and 3, and also shown in Fig. 11 for a graphical representation.

From the presented detection results, we can directly answers the questions posed in the Introduction.

Table 1 Sensor configuration and associated data source

No	Data source	Sample dimension	NN input dimension
1#	All dual-foot data	12	612
2#	All single-foot data	6	306
3#	All dual-foot acceleration	6	306
4#	All single-foot acceleration	3	153
5#	All dual-foot angular rate	6	306
6#	All single-foot angular rate	3	153
7#	Dual-pitch-axis angular rate	2	102
8#	Single-pitch-axis angular rate	1	51

Table 2 Detection sensitivity of different sensor configurations

HMM state	Gait phase	1#	2#	3#	4#	5#	6#	7#	8#
s_1	HS-FF	93.91	94.29	86.63	79.88	94.18	94.34	88.03	86.01
s_2	FF-MSt	88.91	62.34	82.28	31.07	90.04	60.95	72.93	16.73
s_3	MSt-HO	93.42	84.56	87.27	79.17	91.27	84.37	93.11	91.31
s_4	HO-TO	91.25	92.94	83.92	82.87	93.33	95.45	94.18	94.99
s_5	TO-MSw	91.16	88.11	83.20	81.03	89.17	85.05	80.37	74.99
s_6	MSw-HS	99.52	99.61	93.95	87.69	97.79	97.79	97.08	97.62
Average values		**93.03**	86.97	86.21	**73.62**	**92.63**	86.32	87.62	76.94

Table 3 Detection specificity of different sensor configurations

HMM state	Gait phase	1#	2#	3#	4#	5#	6#	7#	8#
s_1	HS-FF	98.93	98.91	97.77	94.52	99.24	99.16	98.41	98.87
s_2	FF-MSt	98.62	96.74	97.34	97.05	98.13	96.60	97.44	97.79
s_3	MSt-HO	97.40	93.86	96.30	90.02	97.90	93.64	96.31	87.64
s_4	HO-TO	97.63	98.01	96.42	95.36	97.85	98.20	98.18	97.97
s_5	TO-MSw	98.87	98.65	97.20	97.56	98.21	98.44	98.28	98.20
s_6	MSw-HS	98.90	98.88	97.49	95.84	98.45	98.35	97.01	96.39
Average values		**98.39**	97.51	97.09	**95.06**	**98.30**	97.40	97.60	96.15

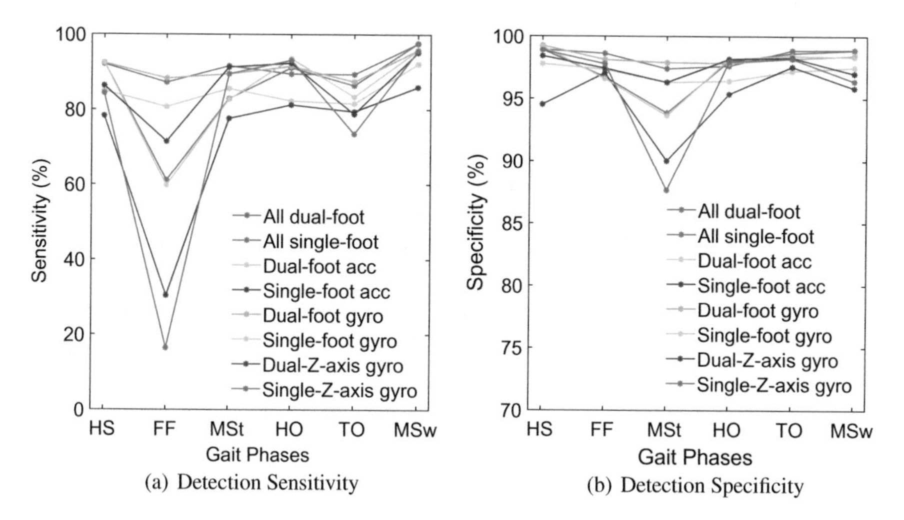

(a) Detection Sensitivity (b) Detection Specificity

Fig. 11 Effect of sensor configuration on the detection of six gait phases

(1) When employing foot-mounted inertial sensors for gait detection during level walking, the averaged performance values of sensitivity and specificity are up to 93.03% and 98.39% respectively, which are calculated with a tolerated timing error less than 2.5 ms by using all available data sources, i.e., all inertial measurements from both feet (1#);
(2) Detection using acceleration from single foot (4#) has the lowest performance values of sensitivity and specificity, which are 73.62% and 95.06% respectively. On the contrary, detection using all inertial measurements (1#) or all angular rates (5#) from both feet provides the highest performance values, and seem to be the most robust to transitions of gait phases.
(3) Angular rates have more profound influence on detection performance than accelerations. The two kinds of data sources 1# and 5#, which use all measurements of gyroscopes from both feet, produce the highest performance values and basically the same detection performance, where the former one (1#) presents a marginally better performance than the latter (5#).

6.3 Discussion

To summarize, the detection results presented in this paper indicate that angular rate holds the most reliable information for gait detection during forward walking on level ground. Particularly, among the triaxial measurements of gyroscope, the angular rate around pitch-axis (7#, 8#) provide more prominent features for gait analysis. Due to the particularity of foot movement, at least two possible reasons can account for this phenomenon:

(1) Although the gyroscope measurements have large bias drifts, their SNR (signal-to-noise ratio) is higher than that of the accelerometer measurements;
(2) The accelerometer measurements are perturbed by the integrated effects of initial alignment error, gravity disturbance, and accelerometer bias error.

Besides, gait detection using measurements from single foot (2#, 4#, 6#, 8#) cannot fully exploit the relationship between two lower limbs, and hence fail to detect the mid-stance (MSt) event that is defined relative to the contralateral lower limb, i.e., the moment that the contralateral heel reaches its highest clearance or the contralateral knee reaches its maximum flexion angle. This problem is even worse when using the acceleration or pitch-axis angular rate from a single foot (4#, 8#). Generally, gait detection using the measurements of both feet is supposed to be more accurate than using that just of ipsilateral limb.

7 Conclusion and Future Work

Gait analysis systems constructed of wearable inertial sensors can be more easily used in both clinical and home environments, which usually have less requirements for operation, maintenance and environment compared to their current counterparts, and thereby leading to a promising future for quantitative gait analysis. This paper aims to investigate the effect of type, number and placement of inertial sensors on gait detection, so as to offer some suggestions for sensor configuration according to the specific requirements. An adaptive hybrid method is adopted for gait detection, which models human gait with a left-to-right six-state HMM, and employs a three-layer neural network to deal with the raw inertial measurements and feed the HMM with classifications.

Generally, placing one sensor on each foot presents less complexity than on various locations of human body, and allows more precise results than on ipsilateral limb by considering the coupling relationship between lower limbs. However, there are several other modes of human locomotion, such as turning, running, sidestepping, walking backwards, ascending or descending stairs, and travelling with an elevator. In these motion conditions, the detection method that relies only on foot motion data may fail to work properly. In future work, sensors placed on other parts of human body will be investigated for gait analysis, such as shank, thigh, and trunk segments.

Acknowledgements This work was jointly supported by National Natural Science Foundation of China no. 61873044, China Postdoctoral Science Foundation no. 2017M621131, Dalian Science and Technology Innovation Fund no. 2018J12SN077, and Fundamental Research Funds for the Central Universities no. DUT18RC(4)036 and DUT16RC(3)015.

References

1. Analog Devices, ADIS16448 (2019-5-4). http://www.analog.com/en/products/sensors-mems/inertial-measurement-units/adis16448.html
2. Oxford Metrics, Vicon Motion Systems (2019-5-4). https://www.vicon.com/products/camera-systems/vantage
3. Abaid, N., Cappa, P., Palermo, E., Petrarca, M., Porfiri, M.: Gait detection in children with and without hemiplegia using single-axis wearable gyroscopes. PloS ONE **8**(9), e73,152 (2013)
4. Abdulrahim, K., Hide, C., Moore, H., Hill, C.: Aiding MEMS IMU with building heading for indoor pedestrian navigation. In: Ubiquitous Positioning Indoor Navigation and Location Based Service, pp. 1–6 (2010)
5. Abdulrahim, K., Hide, C., Moore, H., Hill, C.: Integrating low cost IMU with building heading in indoor pedestrian navigation. J. Glob. Position. Syst. **10**(1), 30–38 (2011)
6. Ayyappa, E.: Normal human locomotion, Part 1: Basic concepts and terminology. J. Prosthet. Orthot. **9**(1), 10–17 (1997)
7. Choe, N., Zhao, H., Qiu, S., So, Y.: A sensor-to-segment calibration method for motion capture system based on low cost MIMU. Measurement **131**, 490–500 (2019)
8. Evans, R.L., Arvind, D.: Detection of gait phases using orient specks for mobile clinical gait analysis. In: The 11th International Conference on Wearable and Implantable Body Sensor Networks, pp. 149–154 (2014)
9. Fischer, C., Sukumar, P.T., Hazas, M.: Tutorial: implementing a pedestrian tracker using inertial sensors. IEEE Pervasive Comput. **12**(2), 17–27 (2013)
10. Godha, S., Lachapelle, G.: Foot mounted inertial system for pedestrian navigation. Meas. Sci. Technol. **19**(7), 1–9 (2008)
11. Guenterberg, E., Yang, A.Y., Ghasemzadeh, H., Jafari, R., Bajcsy, R., Sastry, S.S.: A method for extracting temporal parameters based on hidden Markov models in body sensor networks with inertial sensors. IEEE Trans. Inf. Technol. Biomed. **13**(6), 1019–1030 (2009)
12. Hecht-Nielsen, R.: Theory of the Backpropagation Neural Network, pp. 65–93. Academic Press (1992)
13. Huang, M.H., Shilling, T., Miller, K.A., Smith, K., LaVictoire, K.: History of falls, gait, balance, and fall risks in older cancer survivors living in the community. Clin. Interv. Aging **10**, 1497 (2015)
14. Li, J., Wang, Z., Wang, J., Zhao, H., Qiu, S., Yang, N., Shi, X.: Inertial sensor-based analysis of equestrian sports between beginner and professional riders under different horse gaits. IEEE Trans. Instrum. Meas. **67**(11), 2692–2704 (2018)
15. Mannini, A., Sabatini, A.M.: Gait phase detection and discrimination between walking-jogging activities using hidden Markov models applied to foot motion data from a gyroscope. Gait Posture **36**(4), 657–661 (2012)
16. Meng, X., Sun, S., Ji, L., Wu, J., Wong, W.: Estimation of center of mass displacement based on gait analysis. In: International Conference on Body Sensor Networks, pp. 150–155 (2011)
17. Morris, R., Hickey, A., Del Din, S., Godfrey, A., Lord, S., Rochester, L.: A model of free-living gait: a factor analysis in parkinson's disease. Gait Posture **52**, 68–71 (2017)
18. Ogiela, M.R., Jain, L.C.: Computational Intelligence Paradigms in Advanced Pattern Classification. Springer, Berlin Heidelberg (2012)
19. Qiu, S., Wang, Z., Zhao, H., Liu, L., Jiang, Y., Fortino, G.: Body sensor network based robust gait analysis: toward clinical and at home use. IEEE Sens. J. (2018)

20. Skog, I., Händel, P., Nilsson, J.O., Rantakokko, J.: Zero-velocity detection—an algorithm evaluation. IEEE Trans. Biomed. Eng. **57**(11), 2657–2666 (2010)
21. Strömbäck, P., Rantakokko, J., Wirkander, S.L., Alexandersson, M., Fors, I., Skog, I., Händel, P.: Foot-mounted inertial navigation and cooperative sensor fusion for indoor positioning. In: Proceedings of the International Technical Meeting of the Institute of Navigation, pp. 89–98 (2010)
22. Wang, J., Wang, Z., Zhao, H., Qiu, S., Li, J.: Using wearable sensors to capture human posture for lumbar movement in competitive swimming. IEEE Trans. Hum. Mach. Syst. **49**(2), 194–205 (2019)
23. Wang, Z., Zhao, H., Qiu, S., Gao, Q.: Stance-phase detection for ZUPT-aided foot-mounted pedestrian navigation system. IEEE/ASME Trans. Mechatron. **20**(6), 3170–3181 (2015)
24. Zhao, H., Wang, Z., Qiu, S., Shen, Y., Zhang, L., Tang, K., Fortino, G.: Heading drift reduction for foot-mounted inertial navigation system via multi-sensor fusion and dual-gait analysis. IEEE Sens. J. (2019)
25. Zhao, H., Wang, Z., Qiu, S., Wang, J., Xu, F., Wang, Z., Shen, Y.: Adaptive gait detection based on foot-mounted inertial sensors and multi-sensor fusion. Inf. Fusion **52**, 157–166 (2019)

Author Index

R. Lee (ed.), *Big Data, Cloud Computing, and Data Science Engineering*,
Studies in Computational Intelligence 844,
https://doi.org/10.1007/978-3-030-24405-7

Printed in the United States
By Bookmasters